Aesthetic Practices in African Tourism

Aesthetic Practices in African Tourism explores "Rastahood," a community, youth culture, and new tourist art form created by young men on the margins of the Ghanaian economy as they came of age at the turn of the millennium.

This book focuses on art, music, and affective experience created within tourism contexts, which enabled young men without educational or class capital to achieve mobility through work with foreigners, transforming the temporal horizon by expanding the geographic one. It traces the path that led young men down the path to Rastahood and investigates how they created an art form in, and of, a particular place and then used it to propel themselves far beyond its confines. The book ends with a leap forward into the present, out of Ghana, and beyond Rastahood, as men, now in middle age, look back upon the path that Rastahood created. It explores the social effects of neoliberal capitalism, specifically the rise of neoliberal subjectivities, collectivities, and socialities.

The book will be of interest to researchers in the fields of anthropology, cultural studies, tourism, art, African and Africana Studies, popular culture; gender studies; migration; youth studies and those interested in African cities.

Ruti Talmor is an Associate Professor of Media Studies at Pitzer College and Chair of the Intercollegiate Media Studies Program at the Claremont Colleges. As a cultural anthropologist, an art curator, and a Professor of Media Studies, Talmor's interdisciplinary work centers on how people use aesthetic objects and practices to craft a place for themselves in the world. This diverse but interrelated body of work sits at the intersection of the anthropology of art, media, and visual culture; the scholarship on migration, mobility, and global capitalism; gender and sexuality studies; and critical curatorial practice. Talmor has been a Fellow of the Getty Foundation, the Wenner-Gren Foundation for Anthropological Research, the Social Science Research Council, the Mellon Foundation, the McCracken Foundation, and the University of Michigan's Center for Afroamerican and African Studies.

Routledge Advances in Tourism and Anthropology: People, Place and World

Series Editors:

Dr Catherine Palmer *(University of Brighton, UK) C.Palmer3@brighton.ac.uk*

Dr Jo-Anne Lester *(University of Brighton, UK) J.Lester@brighton.ac.uk*

To discuss any ideas for the series please contact Faye Leerink, Commissioning Editor: faye.leerink@tandf.co.uk or the Series Editors.

This series draws inspiration from anthropology's overarching aim to explore and better understand the human condition in all its fascinating diversity. It seeks to expand the intellectual landscape of anthropology and tourism in relation to how we understand the experience of being human, providing critical inquiry into the spaces, places, and lives in which tourism unfolds. Contributions to the series will consider how such spaces are embodied, imagined, constructed, experienced, memorialized, and contested. The series provides a forum for cutting-edge research and innovative thinking from tourism, anthropology, and related disciplines such as philosophy, history, sociology, geography, cultural studies, architecture, the arts, and feminist studies.

Tourism and Indigenous Heritage in Latin America
As Observed through Mexico's Magical Village Cuetzalan
Casper Jacobsen

Tourism and Embodiment
Edited by Catherine Palmer and Hazel Andrews

Tourism Encounters and Imaginaries: The Front and Back Stage of Tourism Performance
Edited by Frances Julia Riemer

Folklore, People and Place
International Perspectives on Tourism and Tradition in Storied Places
Edited by Jack Hunter and Rachael Ironside

For more information about this series please visit: www.routledge.com/Routledge-Advances-in-Tourism-and-Anthropology/book-series/RATA

Aesthetic Practices in African Tourism

Ruti Talmor

Routledge
Taylor & Francis Group

LONDON AND NEW YORK

First published 2024
by Routledge
4 Park Square, Milton Park, Abingdon, Oxon OX14 4RN

and by Routledge
605 Third Avenue, New York, NY 10158

Routledge is an imprint of the Taylor & Francis Group, an informa business

© 2024 Ruti Talmor

All drawings made by Nicholas Tettey Wayo, based on photographs by Ruti Talmor

British Library Cataloguing-in-Publication Data
A catalogue record for this book is available from the British Library

Library of Congress Cataloging-in-Publication Data
Names: Talmor, Ruti, author.
Title: Aesthetic practices in African tourism / Ruti Talmor.
Description: Abingdon, Oxon ; New York, NY : Routledge, 2024. |
Includes bibliographical references and index.
Identifiers: LCCN 2023035884 (print) | LCCN 2023035885 (ebook) |
ISBN 9780367199920 (hardback) | ISBN 9781032656205 (paperback) |
ISBN 9780429244568 (ebook)
Subjects: LCSH: Tourism--Social aspects--Ghana. |
Heritage tourism--Ghana. | Artisans--Ghana. | Rastafarians--Ghana.
Classification: LCC G155.G4 T35 2024 (print) | LCC G155.G4 (ebook) |
DDC 338.4/791667--dc23/eng/20231018
LC record available at https://lccn.loc.gov/2023035884
LC ebook record available at https://lccn.loc.gov/2023035885

ISBN: 978-0-367-19992-0 (hbk)
ISBN: 978-1-032-65620-5 (pbk)
ISBN: 978-0-429-24456-8 (ebk)

DOI: 10.4324/9780429244568

Typeset in Times New Roman
by KnowledgeWorks Global Ltd.

for Alero

Contents

Acknowledgments

First and foremost, I thank all those whose words and lessons fill this book, for their knowledge, wisdom, generosity, and, in some cases, decades of friendship. Anthropological work could never take place if it were not for those people who open doors to worlds, welcome you in, introduce you to others, and guide you through. The Management of the Greater Accra Regional Centre for National Culture graciously agreed to and supported this study, in particular then-Director Alex Sefah Twerefour. To the members of the Handicraft Market Traders Association, the Handicraft Producers Association, the Ghana Association of Visual Artists (GAVA), the Foundation for Contemporary Arts Ghana (FCA), and the International Centre for African Music and Dance (ICAMD), thank you for your welcome, for the many moments of pleasure, and for ongoing lessons throughout the research. Interviews with over 60 people and conversations with countless others inform these pages, and I wish I could thank them all. My deepest gratitude goes to the Rasta artists I have called Moses Danquah, Samson Donkor, Noah Akarma, Gideon Namoale, Jeremiah Nachinsi, and Ezekiel Poku; to the carvers and traders Godwin Ametewe, Mawuli Awoonor, Afrane Dawson, and Al Haji Mohammed Lawan; and to Angel, Brooke, Jane, Molly, and Rose. So many others in Ghana made this book possible. Special thanks go to Mohammed ben Abdallah, Nat Nunoo Amartofio, Kwadwo Ani, Samuel Ashong, Kofi Gaele Dawson, Rocky Dawuni, Eric Don Arthur, Akwele Suma Glory, Ablade Glover, J.H. Kwabena Nketia, Joe Nkrumah, Alero Olympio, Reggie Rockstone, Virginia Ryan, and Carrie Sullivan.

This project began as a dissertation at New York University's Department of Anthropology, where my training was funded by a MacCracken Graduate Student Fellowship and my research was funded by the Wenner-Gren Foundation's Dissertation Fieldwork Grant 7058 and the Social Science Research Council's International Dissertation Field Research Fellowship. I thank my three dissertation advisors, Thomas O. Beidelman, Fred Myers, and Steven Feld. Each of them has shaped a particular part of my mind, and each time I use it I do so in dialogue with them. I owe a similar debt to Faye Ginsburg, Bambi Schieffelin, and Thomas Abercrombie. Thank you to Deborah Pellow for her brilliant writing and her personal generosity and support.

The project grew at the Center for Afroamerican and African Studies at the University of Michigan, where I was a Du Bois-Mandela-Rodney Postdoctoral Fellow. A special thanks to Kevin Gaines, Kelly Askew, David Doris, Elisha P. Renne, Raymond Silverman, and the inimitable Julius S. Scott. It grew again at the John B. Hurford '60 Humanities Center at Haverford College, where I held a Mellon Postdoctoral Fellowship. My deepest thanks to Kimberly Benston, then director of the Center, to Ken Kolton-Fromm, who created and led the "Material Identity" seminar, and to its faculty members Laura McGrane, Hank Glassman, Darin Hayton, Jesse Weaver Shipley, and Travis Zadeh. Beyond the seminar, thank you to Laurie Kane Hart, Philippe Bourgois, Vicky Funari, John Muse, Matthew Callinan, James Weissinger, Emily Cronin, and Hilary Parsons Dick. My time as a Getty Research Scholar during the Getty Research Institute's "Art and Anthropology: The Agency of Objects" theme year was invaluable: a special thank you to Alexa Sekyra and David Brafman, and to my fellow scholars Katie Scott, Lyneise Williams, Julia Christine Lum, and Zirwat Chowdhury. I first worked through the ideas in Chapter 3 in an article titled "Masks, Elephants, and Djembe Drums: Craft as Historical Experience in Ghana" (2012) in the *Journal of Modern Craft*, and I am grateful to editor Glenn Adamson for the invitation, support, and suggestions which greatly improved the piece. The ideas covered in Chapter 5 were honed by my participation in the invited workshop "Love in Africa," organized by Altaïr Despres, Jennifer Cole, and Lynn M. Thomas. Much of this book was written at the Huntington Library due to the generosity of its Readers Program.

Nicholas Tettey Wayo created the five graceful drawings in the book based on my photographs. Jaida Samudra and Jay Fischer provided essential editorial skills. Christian Lopez helped tackle bibliographic logistics.

I am deeply fortunate to work at Pitzer College, a member of the Claremont Colleges, where I have received much support for this book, including yearly research funds, sabbaticals, a semester as Scholar-in-Residence, the freedom to teach widely and creatively, and students who regularly ask brilliant questions. To my wonderful colleagues in Media Studies, Elizabeth Affuso, Alexandra Juhasz, Gina Lamb, Jesse Lerner, Ming-Yuen Ma, Stephanie Hutin, and Eddie Gonzalez, thank you for your support, friendship, and inspiration.

Harmony O'Rourke, thank you for your close and generous reading and endless conversations about Africa. To Leila Barratt Denyer, Michelle Berenfeld, Jen Berkowitz, Rene Gerrets, Carina Johnson, Devon Archer Keen, Avi Lanczos, Marian Maloney, Ana Martins, Delali Noviewoo, Susan Phillips, Laurel Rose, Michael Shaw, Sabra Gayle Thorner, Patricia van Leeuwaarde Moonsammy, and Carlin Wing, and to Harmony again, thank you for the conversations, the meals, the laughter, and the constancy of friendship. A special call out to two dear friends without whom this book is unimaginable: Jesse Weaver Shipley, who has been my generous interlocutor since our first encounter, when I was preparing for fieldwork in Ghana, and to Susanna Rosenbaum, with whom I think about everything.

To my parents, Lihie and Mishael, and my sister Dafna, this project, like much of our individual work, is born of the movements and resultant longings and obsessions that have defined our family, as well as our constant choice to process these in artistic ways. Augie Isidore, you make everything possible—thank you for your love and support, your insight and humor. Isaiah Isidore Talmor, you are the son and the sun.

Introduction

I and I: Artmaking, Mobility, and Intercultural Reproduction

In 2012, Roskilde, Denmark, Ghanaian drummer Moses Danquah steps out of a crowd of scantily dressed White Europeans onto the enormous campground of the music festival. His drum, a *jembe*, is suspended from two red straps crossed over his chest. The rawhide drumhead faces out, the foot of its goblet-shaped hardwood shell receding from view between the drummer's legs. He wears a black T-shirt and jeans over his long-muscled body, new Nike sneakers, a red, green, and yellow wristband, and a single cowrie shell suspended from a silver chain. He periodically swings shoulder-length dreadlocks away from his face. People mill around him, drinking beer.

Moses bends lightly at the knees and leans forward over the drum. He smiles and begins to play. The crisp sounds of the jembe call out. The crowd turns toward the drumming, creating a ring of space around Moses. Three other drummers, old friends arrayed behind him, answer the call and begin to play their own drums. The jembe seem to talk. The crowd begins swaying to the rhythm; some listeners begin to dance.

Moses' smile broadens. He mimes a conversation with the audience between beats. He breaks, suddenly silent and still, legs landing wide apart, then points to individual drummers in his group. He twirls, the drum suspended before him, lunges toward the audience, bending deeper, leaning closer, his every motion sharp and loose. He stops dancing—now only his hands are flying, a blur on the rim of the drum, the perfectly timed slaps marking out a bright, high-pitched rhythm, every muscle on his arms precisely articulated. Still smiling, he is completely focused on the music. He turns his head, looking into the distance as he plays. His dreadlocks quiver and the cowrie shell shines, flashing bright white against the black t-shirt.

With a dramatic flourish, he is done. There is a moment of echoing silence. Then the crowd erupts in cheers and applause.

***

I first met Moses in a fishing village outside of Cape Coast, Ghana, in 2001, but I did not see him and his drumming and dancing troupe, Black Star, perform until a few months after we met, when they traveled from Accra to the city of Cape

DOI: 10.4324/9780429244568-1

Coast to perform at PANAFEST. First created by Ghanaian President J.J. Rawlings, PANAFEST is a Pan-African cultural festival held intermittently in Ghana since it was founded in 1992. Co-sponsored by the African Union and the Government of Ghana, with the support of international non-governmental organizations and the private sector, PANAFEST welcomed people from the African diaspora back to Ghana through performances, concerts, and ritual events. Such cultural activities are often held at the infamous former slave forts of Cape Coast and Elmina, important sites from the cross-Atlantic slave trade that have become key destinations for diasporic memorial tourism. PANAFEST has provided an excellent opportunity to take in live performances from throughout Africa and the diaspora.

When I first saw Black Star, they were engaged in backstage preparations, readying themselves to perform their blend of West African rhythms and dance acrobatics at Oasis, a local Cape Coast spot frequented by Ghanaians and foreigners. At a ramshackle house on a cliff behind the slave fort, the entire band was outfitting themselves in what might be called neotraditional Ghanaian garb.[1] They wore matching outfits of brightly patterned waxprint and batik cloth, headdresses, and anklets of raffia, and had ritually painted their faces and arms with white chalk.

On stage, the drummers held nothing back as the dancers and acrobats pushed their bodies to the limit. I was captivated by Black Star's performance even as the ethnographer in me clinically observed myself being entranced.[2]

Searching the festival grounds the next day, I expected to find the band in the audience at one of the stages of PANAFEST. Instead, I found them avidly watching a small television set suspended from a high corner of the temporary food court set up for the festival. A VHS tape of a Bob Marley concert was playing. Now dressed in Rastafarian garb instead of neotraditional Ghanaian costume, the members of Black Star and most of the other young men in the restaurant were singing along with Marley. Many knew by heart even what Marley said in between songs and recited his words with him, imitating his Jamaican accent and speech rhythms, voicing their support with interjections of "Yes, Irie—respect—Rastafarai." I was struck by the intense focus on and emotion evoked by this recorded playback of a Rastafarian cultural performance from the diaspora (which clearly everyone had also seen before) in the midst of what is ostensibly Ghana's largest live display of its musical culture, and what is more, that it was some of the very producers of that live festival who chose to spend their time rewatching this mediated, long-ago performance instead of taking in the live music that filled the festival.

By then, I already knew that life for most of Moses' group was precarious. Some days they could barely scrape together the money for a single meal. Everything—from where to sleep to how to travel and transport their drums to Cape Coast—was a logistical and financial challenge to be overcome in some unforeseen way. Having made it to the festival was a collective achievement, a unique opportunity. Yet, given the choice between attending live (and thus inherently unrepeatable) performances and watching a scratched-up replay of a memorized concert video, all the Black Star members had chosen the latter. The contrast between shared liveness and mediated replay, between staged performances of neotraditional Ghanaian culture and everyday Rastafarian performativity, between that which is available nearby and that which is distant and desired, presented a puzzle that stimulated my research into Rastaworld over the next 15 years.[3] This trilogy of scenes—the backstage preparations in the house behind the fort, the staged neotraditional drumming and dancing performance, and the genuine pleasure of consuming Marley at the food court—became my portal into this project. All the components of Rastaworld were there, including my own female *oboroni* (stranger, foreigner, White person) self, even though it took some time before I understood this.

***

Rastahood

Every element of both Black Star's 2001 show and Moses' 2012 performance reveal mastery of what I call Rastahood, a new form of touristic art that developed in Ghana at the turn of the millennium. In response to economic changes that challenged careers in other forms of tourist art, young Ghanaian craftsmen began making and selling jembe drums to tourists, while presenting themselves as Rastafarian-neotraditionalists in everyday life as well as in staged and informal drumming and dance performances. Although not adherents to Rastafari or traditional Ghanaian religions, styling themselves Rasta (as Ghanaians refer to the youth culture) enabled these young men to make a living in the tourist art trade in Accra, Ghana's capital, and in other cities and towns along its coast. In this book, the term "Rasta" always refers to the youth culture, not to followers of the Rastafari religion, although several such religious communities do exist in Ghana (Alleyne 2017, Middleton 2006, Savishinsky 1994a, 1994b, White 2007, 2010). Moreover, I am specifically interested in Rastahood as it developed at the Accra Arts Centre, Ghana's largest center for tourist art. Key to this story of art's intercultural potential is the site of Accra, Ghana's multicultural, cosmopolitan capital, long defined and produced by exchanges and collaborations between "natives" and migrants, and the Accra Arts Centre, a place that, in its productive multiplicity, is a microcosm of the city's potential. Starting here, youth developed the Rasta figure, a postmodern pastiche of decontextualized, recombined, and repurposed elements that eventually allowed them to create a new commodity for global tourism and the tourist art trade, to constitute a Ghanaian masculinity for the 1999 generation (youth coming of age at the turn of the millennium), and to move far beyond Accra.

The question of what exactly Rastas *make* is central to this book. Unlike other forms of Ghanaian touristic art such as woodcarving, brasswork, or basket making, Rastas do not craft a single, material object. Rastahood is an entire package in itself, combining production of a physical souvenir object (the jembe drum), somatic stylization (through fashion, style, language, gesture, and everyday performativity), staged performances of drumming and dancing, curation of the touristic experience, and various forms of affective labor (including romance).

From jembe and dreadlocks to cowrie shells and Nikes, young Ghanaian Rastas at the turn of the millennium materialized the forces that were shaping their lives: the global circulation of West African drumming and Jamaican reggae music; Rastafarianism as cosmopolitan ideology of resistance, redemption, and unity; postcolonial commodification of ethnic practices for the tourism market; and late capitalism's transformation of labor, culture, care, and time. Hence, the boundaries of Rastahood's objects and the traditions that engendered them are multiple, as Rastahood appropriated and merged three cultural repertoires into a single form to meet the existential needs of a generation of Ghanaian youth and their foreign counterparts. I view Rastas not as instrument makers, performers, or hustlers, but rather as multimedia artists and translators of foreign image and desire into form and experience, through which Rastas could engage both mainstream Ghanaians and foreign tourists to craft alternative paths to a good life when previous ones had been foreclosed.

In a country full of globally renowned artistic forms, from brass to textile to wood to music, what Rastas make tends to be demeaned. But following Bakhtin et al. (1990), I see art as a mode of consciousness, of enframing and setting apart, of creating a space around the art object which can then allow for a view—a gaze across a distance, no matter how small—which in turn can engender social, political, cultural, and economic transformation.[4] I see it as an impulse which can take many forms. In this case, what Rastas make is a figure, a term I choose for its capacity to hold the concepts of image, embodiment, performance, and pattern. According to this definition, Rastas make art.

How and why Rastahood arose as an artistic solution to the problems of sustaining life and livelihood under late capitalism in a West African context is thus the subject of this book. We find out that Rastahood is deeply tied to modes of prestige and mobility in Accra craftworlds centered at the Accra Arts Centre, moving out from there into the wider world. Most of the book covers the years 1999–2005, when this version of Rastahood was rendered necessary by neoliberalism, made possible by the opening up of Ghana's public sphere in the latter years of the PNDC regime, and created by the generation that came of age at this moment in time. The book starts by going back in time to trace the roots of Rastahood at the Arts Centre and ends by shifting to the present time, when the teenage boys who became Rastas are now adult fathers, many living abroad. *Aesthetic Practices in African Tourism* does not focus on the handicraft trade or the tourism industry as economic formations. Rather, it traces the path that led young men down the path to Rastahood and investigates how they created an art form in and of a particular place—the Arts Centre—and then used it to propel themselves far beyond its confines. It therefore attends primarily to the experiences, narratives, actions, and relationships of the artists known as Rastas.

The Neoliberal Crisis in Social Reproduction

I situate this book in a body of scholarship that has developed around the social effects of neoliberal capitalism, specifically the rise of neoliberal subjectivities, collectivities, and socialities. First in Africa and then elsewhere, this literature initially focused on the crisis of social reproduction and the plight of "youth," the many young people who have not been able to acquire the resources necessary to fund a marriage and attain full social status as adults (Cole 2010, Diouf 2003, Durham 2000, Gable 2000, Honwana 2012, Honwana and De Boeck 2005). Subsequently, this work focused on the unanticipated effects of decentralization on collective identities—the weakening of national identities and emergence of sub-state and globalized collectivities (Ferguson 2006, Piot 2010).

Every generation encounters the world anew and has a new "experience of the present," a new "structure of feeling" which produces a practical conscious-ness, "an embryonic phase of thinking and feeling before it becomes fully articu-late and defined exchange" (Williams 1977: 128–132). In Ghana, as elsewhere, people experienced what Berlant calls "aspirational normativity" (Berlant 2011). They longed for the normative life course, with its gendered models of consumer-ism, reproductive futurism, and naturalized shifts from child to adult, dependent to supporter, and client to patron (Edelman 2004). However, the achievement of adulthood was impeded by the increasing precarity of labor seeping into other do-mains of life (Allison 2013, Berlant 2011, Comaroff & Comaroff 2001, Neilson and Rossiter 2008, Povinelli 2011). Consequently, as a category of personhood, "youth" expanded, as the economic resources needed to move into adulthood be-came harder to access (Cole & Durham 2007, Durham 2000). This meant that as a temporal category, "youth" shifted from being a brief transition between childhood and adulthood to becoming a bewilderingly amorphous, ongoing present, a period of "waithood" upon which youth comment and act in different ways (Cole 2010, Dhillon & Yousef 2009, Honwana 2012, Honwana & de Boeck 2005, Mains 2007, Masquelier 2013, Quayson 2014, Ralph 2008, Singerman et al. 2007, Sommers 2012, Stasik et al. 2020, Weiss 2004). Standing on corners, sitting on verandas, resting in parental living rooms, youth in waiting became embodiments of a larger crisis—an inability to see the future and experience time in an orderly, progressive fashion.

Lack of employment prevented economic independence and thus trapped sub-jects in a protracted state of youth, the transitional category between dependent childhood and independent adulthood. The terms "waithood" (first introduced by Diane Singerman), "a generation in waiting," and "wait adulthood" have come to define this (often gendered male) state of prolonged sociocultural and economic limbo (Honwana 2012, Singerman et al. 2007, Sommers 2012, but see Moore 2006 for women). While this "neither-here-nor-there state" (Honwana 2012:3) is an in-creasingly shared millennial condition that is merging the destinies of the Global North and South, it had already defined the experience of youth in the Global South for several decades (Allison 2013, Auyero 2012, Comaroff & Comaroff 2012, Mains 2007, Vigh 2009, Wacquant 2008).

This crisis in social reproduction shaped not only personal experiences but also the relationships between generations and between women and men, as individual life scripts and assumed social contracts between these groups could no longer be fulfilled (Christiansen et al. 2006, Coe 2011, Comaroff & Comaroff 2001). Older and younger men and women all found themselves "caught in a seemingly unbridgeable schism between the culturally expected and the socially possible" (Vigh 2009:95–96). Such moments of cleavage from older social schema (Johnson-Hanks 2007; Kasfir and Förster 2013:13) resulted in youths turning their imaginations to other places (Appadurai 1991, Cole 2010, Ferguson 2006, Lucht 2011, Vigh 2009). Indeed, trapped in a "culture of masculine waiting" (Jeffrey 2010), increasing numbers of young men from throughout the Global South would come to view migration and wage labor in the Global North as the best (perhaps only) way to achieve a successful livelihood and adulthood at home.

In Accra, the city where much of this book takes place, one finds the *kòbòlò* (Ga. Pl. *kòbòlòi*), the "good-for-nothing street lounger" (Quayson 2014:199), a by now familiar figure embodying "a transitional state of urban existence at the intersecting vectors of space, time, and longing," a "masculine sphere of crisis," and "a structural transition between socially acceptable age-related activities" (2014:199). For English scholar Ato Quayson, the *kòbòlò* is a liminal figure, in between life phases, in between precarious, low-paying jobs, often looking for a way out of the country as a way out of this phase of waithood.

As Ghanaian youth would say, "There is no money in the system." Geographic mobility became linked with a "symbolic 'moving up,' be it economic, social, or cultural" (Salazar & Jayaram 2016:2). Thus, Ghanaian youth gazed outward in their search for an adult future. And yet, while migration had become the best hope for economic mobility, it was fraught with difficulty, from the financial hurdles and dangers of travel itself to the challenges faced upon arrival in a foreign country. Following the rise of the neoliberal economy in Ghana in the 1980s and the country's re-opening as the "Gateway to Africa" in the early 1990s, globally engaged art and music careers offered an alternative to both manual and dwindling, postgraduate labor (Feld 2012, Herz-Jakoby 2013, Osumare 2012, Schauert 2015, Shipley 2013). This book focuses on art, music, and affective experience created within tourism contexts, which enabled young men without educational or class capital to achieve mobility through work with foreigners, transforming the temporal horizon by expanding the geographic one.

Artworlds as Contact Zones

Of course, this is an instance of a broader phenomenon. African artworlds have long been contact zones between artists seeking a new market and patronage base and foreigners interested in returning home with pieces of the African continent for souvenirs (Bascom 1976, Ben-Amos 1976, 1977, Jules-Rosette 1984, Kasfir 2007, Pratt 1992, Price 1989, Sandelowsky 1976, Steiner 1994). As Myers (2002) explains in the context of Australian Aboriginal artworlds, the market is not a separate domain but a structure of symbolic transformation; the market regime does not replace other regimes of value; it reorganizes them, existing, as it were, as a kind of zone where they come into contact. Pratt's original concept of the "contact zone" refers to "social spaces where cultures, meet, clash and grapple with each other, often in contexts of highly asymmetrical relations of power, such as colonialism, slavery, or their aftermaths as they lived out in many parts of the world today" (Pratt 1992:34, see Mudimbe 1988). The production and exchange of craft in Africa has historically been both a product and productive of contact zones, in which artists from different ethnic groups encountered Westerners and each other (Clifford 1997, Schildkrout & Keim 1998). I follow William Pietz in viewing these zones as "inhabited intercultural spaces ... whose function was to translate and transvalue objects between radically different social systems" (Pietz 1985:6). In turn, the objects made within them became "fetishes" that acted "as a material space gathering an otherwise unconnected multiplicity into the unity of its enduring singularity" (Pietz 1985:10, 15). In this way, contact zones made art, and art made sense of contact.

The elements of asymmetry and contact are essential to my use of the term and similar terms such as "interculture" and "intercultural." I follow Abercrombie (1998) in treating the interculture as a space of contact critically defined by unequal power relations between two or more participants. In African intercultural zones, artmaking has often involved a double act of translation, as foreigners sought objects that would explain and materialize the realities they confronted in Africa, while African artists sought to understand and transmute those foreign visions into form. The Accra tourist trade, in which the 1999 generation came of age, had been such a contact zone since the colonial period, when handicraft carvers and dealers collaboratively translated British desires for African souvenirs into wood, a metamorphosis that enabled Northerners and Ewes, both marginalized strangers, to make a living in Accra.[5]

Much like the older generations, Ghanaian Rastas entered this transactional space in search of a better life, and, like their predecessors, they produced a new art form made possible and necessary by the socio-economic conditions of the new millennium. This art form—the Rasta figure—combined body art, performance art, and object-making and merged the appeal of the African drummer with the allure of the Rastafarian. Inserting this figure into neoliberal economies of care, desire, and experience, Ghanaian Rastas created, or arguably became, a portal for tourists into a desired, imagined "Africa" and, for themselves, into a desired, imagined "abroad."

Diasporic Resources

Obviously, the Rasta figure presented to this gaze draws heavily on diasporic culture. The uptake of reggae and Rastafarian imagery, language, and culture, discussed in Chapter 4, are unsurprising, as these became hegemonic forms of globally circulating Black culture in the 1980s, taken up by youth around the world due both to its dominant commercial appeal and its potential for resistance. But as I argue in Chapter 3, Rastas' uptake of the Sahelian jembe drum can also be seen through a diasporic lens, since the drum had to gain power in the diaspora before it could serve its function in Rastahood.

In combining these non-Ghanaian forms with national musical cultures, Rastas engaged in a "re-remixing of forms" (Clarke and Thomas 2006:26) that defines the African diaspora as an ongoing process of relation and transformation (Gilroy

1987, 1993, Hall 1993, see Campt 2005, 2012). Anthropologist Jacqueline Nassy Brown has coined the term "diasporic resources" to describe "those practices in which black people ... make use of any of the vast resources of what they construct as the black world, yet within the political economy of what has been available to them. Diasporic resources may include not just cultural productions such as music, but also people and places, as well as iconography, ideas, and ideologies associated with them" (Brown 1998: 297).

For Brown, localized appropriations of diasporic resources are always done "to meet particular needs—but do so within limits, within and against power asymmetries, and with political consequences" (Brown 1998: 297). Within this framework, even though they contended with accusations of imitation and inauthenticity, Ghanaian youth could access diasporic resources because these in some sense belonged to them through shared origin. They could then use these in political projects of emancipation both nationally and internationally (Gilroy 1987). To reach a global audience, Rastas needed globally recognizable forms, and they found these in the musical cultures of reggae and jembe rhythms. In this appropriation of diasporic, continental, and global resources, Rastas were following in the footsteps of their elders, creating a new art form to respond to and transform the conditions of the present, and did so alongside other Ghanaian youth, who localized diasporic forms, including hip hop (Osumare 2012, Shipley 2013) and Jamaican dancehall (Alleyne 2017) in similar ways.

"A Beckoning Elsewhere": Rastas, *Aborofo*, and the Making of *Nkabɔm*

I borrow the words of photographer and writer Stanley Wolukau-Wanambwa to ask, "[F]or whom, or *to* whom, is the performance addressed, and what does it bring into immediate being?" (Blight and Roediger 2019:187). How does the Rasta figure "come into visibility" (2019:181)?

As a form, Rastahood is simultaneously individual and collective, its authorship at times as slippery as its objecthood. It merges two globally circulating representations of the continent, one diasporic—Rastafarianism (originally from Jamaica)—and one African but globalized through world music—jembe rhythms (originally from Guinea). These representations are localized, or indigenized (Osumare 2012), with the addition of various Ghanaian drum rhythms, clothing, and ornamentation, as Rastas perform a neotraditional identity that was not the cultural property of any single ethnic group. Most Rastas in this study identified with one of the four largest ethnic groups from the south—Ga, Fante, Ashanti, or Ewe—but collectively appropriated drum sets, rhythms, concepts, and names from all four. A smaller group of Rastas were Northerners, but again, elements from that region were borrowed by all. Rastahood thus merged (1) Rastafarian signs (including dreadlocks, accessories in the Rasta colors, and a spoken and gestural vocabulary); (2) neotraditional, pan-ethnic signs (such as jewelry and accessories containing cowrie shells, beads, and bone, and various types of West African cloth); and (3)Afrocentric clothing and accessories (featuring iconography such as lions, drums, or village scenes).

Many of these were images of "Africa" made from without or co-constructed by Ghanaians and foreigners. This bricolage created a polysemic figure (Hebdige 1979)—the Rasta—both anchored in locality and traditional culture and plugged into global modes of resistance and cosmopolitanism.

However, while it drew heavily upon diasporic signs and symbols, Rastahood's primary interlocutors were White and from Western Europe or North America. In local parlance, they were *aborofo* (Akan: singular *oboroni*; strangers from *aborokyire*: the land beyond the horizon).[6] The terms oboroni/aborofo are the de facto Southern Ghanaian terms for foreigners. They usually mark differences in wealth, culture, and race between people of Ghana and people from the Global North (Bosiwah et al. 2015). Race is the most complicated aspect of these terms, since the terms are sometimes used to refer to light-skinned foreigners from Latin America and the Middle East, people from the African diaspora, and even light-skinned Ghanaians (Bosiwah et al. 2015, Hartman 2008, Holsey 2008, Pierre 2012). Instead, the term transcends corporeal race to align Whiteness with wealth, status, privilege, and property (Harris 1993, see Dyer 1997).[7]

I use the terms "oboroni" and "aborofo" throughout this book because they connote a localized ensemble of phenomena (Frankenburg 1997), presenting the world with Ghana at the center. Moreover, the oboroni/aborofo terms are shaped by an imagination of the other, a desire for that which is lacking in oneself or one's place and which is projected onto the other. As such, the imagined oboroni subject mirrors the imagined Rasta subject. The terms linguistically reproduce the boundaries of the intersubjective space of Rastaworld, the zone where Rastas and aborofo come together in search of an elsewhere, and out of which they emerge having produced new realities through one another.

By becoming artists who represented Ghana and Africa to foreign eyes, Rastas created Rastaworld and engaged in transformative exchanges with foreigners, initiating a new kind of contact with the world outside Ghana and enabling them to access aborokyire. Like the related terms aborofo/oboroni, the root term *aborokyire* indexes a Ghanaian theory of globalization, which obviates the problematic partiality of terms such as "West," "Global North," or "developed world" (see Salazar 2010). An idealization of abroad as a "source of development and means of escape" (Slater & Kwami 2005) inheres in the term, which mythologizes "a 'white,' non-African elsewhere" (Ferguson 2006: 154) as a magical site of redemption and "the horizon of hope in the imagination of the postcolonial subject" (Asempasah & Sam 2016).

Therefore, I view Rastas and aborofo as parallel embodiments of the other's desire—bodies that bracket and produce the contact zone of Rastaworld. Similarly, aborokyire, as a spatialization of said longing, is paralleled by a utopian "Africa" that existed alongside the persistent, negative image of the content in the Western imaginary. This utopian Africa was a mirror made up of negative spaces in which the oboroni subject sought itself. It was, in the words of Paulla Ebron, "evocative, disruptive ... the outsider continent, the recalcitrant space, the resistant thorn that pricks the sides of the West in its refusal to embrace 'the modern project.' ... Africa becomes a countermove, a trickster, a hero figure that can combat the disaffected sentiments of the iron cage of modernity" (Ebron 2002:2).

Crucially, it was in Africa's "culture," its "traditions," that the escape from the iron cage could be found. By the late 1990s, this image of Africa had proliferated in Western and diasporic media and brought tourists to Ghana (Mbembe 2001, Mudimbe 1988, Salazar 2010, Sheller & Urry 2004). Rastahood could not have existed without it. For many aborofo, the journey to Ghana was a quest for something lacking in their lives back home.[8] They found in Rastaworld precisely what they searched for. Ghanaian youth fashioned themselves into a new image of "Africanicity" (see Barthes & Heath 1978)—what a Gambian interlocutor of anthropologist Paulla Ebron termed "the Africa" (Ebron 2002:3), and specifically, a Black African masculinity that spoke to oboroni desire by drawing upon a globally circulating continental icon—the jembe drummer—and a diasporic image—the Rastaman.[9] Rastahood, then, was produced by and produced constructions of both Blackness and Whiteness, often times as opposite poles, of Africa and aborokyire as the "beckoning elsewhere" (Gordimer 2007:5) that pulled both parties into Rastaworld in the first place. Both Rastas and aborofo wish to cross over into the space of other, both spaces envisioned as sites of escape from conditions of selfhood at home. Thus framed, Rastaworld became a Ghanaian field of intercultural production, a contact zone defined by a distinct set of rules, aesthetics, and possibilities that cut through Accra, other parts of Ghana, and, finally, aborokyire.

The term *nkabɔm* (Twi: unity) similarly defines the space of Rastaworld and aspirations of Rastahood. A call for nkabɔm was one of the most enduring and oft-repeated messages I heard from Rastas. They sometimes glossed nkabɔm as "all of us together still." The "all of us" called into being was global and intercultural; the desired unity was one that would produce *communitas* (Turner 1969), transcending boundaries and addressing inequities between Ghanaians and aborofo that had past colonial roots and continued in globalized present. "All of us together still" could build a new world, with each contributing their strengths by engaging in collaboration and fair exchange. Rastahood attempted to produce this world. This was the intercultural, intersubjective space shared by Moses and his audience in Denmark in 2012, by Black Star and theirs in Ghana in 2001. That space, and Moses' act of crossing it with his drum, is the subject of this book.

The Rasta Figure

For Rastas, the central *material* object of exchange was the jembe, a goblet-shaped drum hailing from nearby Guinea and Mali that had been transformed by global flows of African music into a potent, highly desirable icon of the continent. It was the crucial item for their engagement in touristic arts as a mode of intercultural exchange, as people translate places into consumable objects, then reach across national, cultural, and linguistic boundaries to hand over these objects in exchange for foreign capital (Marcus & Myers 1995, Phillips & Steiner 1999). The jembe thus became the most popular drum in the Ghanaian tourist trade, becoming not merely or mostly musical instruments for many of the young tourists who bought them. They were tangible evidence of Ghana/Africa, "mementos and proof of having been there" (Fabian 1998:120, Phillips 2002), found in the corners of many rooms the world over as placeholders and conversation-starters. The object as souvenir became a repository for memory of an entire experience in Ghana.

But the jembe was only part of a total figure that called out to tourists, addressed them with a promise, embodying their desire. Buying the drum was not simply the purchase of an object; it was advance payment for a series of intense encounters that often began with euphoric experiences in Rasta productions, including drumming and dancing classes, jam sessions, and formal performances at music venues and cultural festivals. Moreover, both aborofo and Rastas hoped such experiences would transcend touristic superficiality. They might be followed by Rastas guiding tourists out of touristic zones of experience into sites of "authenticity" such as rural villages and working-class urban neighborhoods and into friendships, flirtations, sex, or romance. Rastas opened doors to Ghana by acting as informal, unpaid tour guides, travel planners, fixers, and linguistic and cultural translators. They used their knowledge, called in favors, arranged contacts and contracts, and mediated between aborofo and Ghanaians. Rasta art thus exemplifies the immaterial yet corporeal affective labor so essential to the global economy today. When tourists purchased jembe drums from Rastas, instead of closing off the experience by obtaining a souvenir, they engaged in a kind of futurity. The object was no longer an encapsulation but a beginning, a portal to new experience for both members of the exchange. Purchase of the drum was a gateway to an art form comprising experiences, produced by Rastas, collected by aborofo.

Gender was a key component of this art form: the Rasta described in this book is always male and read through the lens of a sexualized, racialized Black masculinity appealing to foreigners. While Rastas often interacted and collaborated with oboroni men, romantic relationships with oboroni women were very common. In fact, many Ghanaians, when asked what defined Rastas, said that they wanted "to catch a tourist," an oboroni woman who would help them reach aborokyire in exchange for being cared for by Rastas while in Ghana. As friends, lovers, guides, or teachers, Rastas worked to bridge the distance between these women and Ghana.[10]

Where Is Rastaworld? Touristic Borderzones

The Rastaworld I describe bridges the tourist-artworld with that of tourisms more broadly, sharing traits with leisure, cultural, heritage, and romance tourism as well as volunteerism. Anyone who has traveled to a coastal tourism destination in the Global South has likely encountered someone like a Rasta: a local youth, often dreadlocked, who hangs out where tourists are and tries to make connections with them and between them and the local community. What distinguishes the practice I describe is its inception as artistic action within the Accra Arts Centre, a gathering place for a multiplicity of aesthetic representations of Africanicity, many defined by bricolage that merged the old with the new, the foreign with the local, in response to the conditions of the present.

Over time, as Rastahood became a reason for tourists to visit Ghana, Ghanaians involved in a range of private enterprises and public institutions benefited from and depended on Rasta forms of work, even as the Rastaworld depended on them to instantiate itself. Many of its sites combined a bar, restaurant, dance club, performance stage, drumming and dancing school, and tourist hostel. These locations provided a total touristic experience within a Rastafied tourist bubble embedded within larger cities and towns. Aborofo on vacation from their ordinary lives visited these locations to experience "African culture," adventure, and play.

Concretely, Rastaworld comprised the bars, performance venues, drumming and dancing schools, cultural centers and institutes, beaches, hostels and hotels, restaurants, and all other spaces where Rastas and tourists converged, when they did so. But symbolically, it was always larger than its geography at any given time. Existing as a potentiality and an imaginary as much as a built space, it generated new aestheticized and recognizable sites the world over (Jackson 2001), as well as new routes through that world (Clifford 1997).

Within Rastaworld, Rastas forged connections not only with tourists but also with volunteers, NGO workers, students abroad, and various expatriates who considered Rastahood an appealing version of the Ghanaian experience. Such connections resulted in dramatic pivots in some Rastas' life trajectories, as professional opportunities or personal relationships led them to move to Europe, Canada, and the United States. At the intersection of tourism, tourist art, and world music, Rastaworld became not only a space of reinvention but also an actual route out of one place and into another, as well as a trajectory out of immobility.[11] Rastaworld was thus a "migration infrastructure" (Xiang & Lindquist 2014), both a portal into Ghana and a route to elsewhere.

In entering a venue where a Rasta drumming and dancing group was performing or a "spot" (local slang for bar) where Rastas congregated, aborofo sought tourism not as a blocking out of the world but as an opening up to it, a "reaching out across cultural differences through dialogue, aesthetic enjoyment and respect, of living together with difference" (Werbner 2008:2). They sought a connection with the other that would lead to a transformed experience of the self. The message of Rastaworld, borrowed from Rastafarianism and world music, was a promise of nkabɔm, universal unity through intimate, (inter)cultural production. For foreigners, especially women, Rastaworld could be a transformative space. For Rastas too, intercultural exchange supported a more expansive sense of self and place in the world and offered refuge, or escape, from economic immobility. Contact with the foreign-at-home could open up opportunities for connecting with the foreign-abroad to create a new being-at-home.

And yet, the story of intercultural contact between Rastas and aborofo can also be told as one of exploitation and misrecognition. Rastas experienced and witnessed poverty, exploitation, and lack of access to resources within Ghanaian society and were profoundly aware of the inequalities between Africa and the West. The reality of the circumstances in which they and their families lived led them out of necessity to enter the tourist art trade. Once within the trade, they were repeatedly confronted by the inequality of these circumstances, as they spent time with tourists who in any single hour could spend more money than their families could spend in a week, tourists moreover whose bodies were loaded with wealth in the form of cellphones, cameras, hiking gear, power bars, mineral water bottles, and other expensive accouterments of travel. This profound economic imbalance hence undergirded every exchange, as Rastas responded to the conditions that had produced the contact zone and its concomitant opportunities.

Hence, Rastahood produces a particular way of inhabiting and experiencing Ghana. Touristic borderzones throughout the world are spaces to which modern mobile subjects travel precisely in search of that which has been lost, for if mobility and change are central to modernity, then the imagined tradition of the immobile "other" in both space and time represents stability (Adey 2010, Salazar 2010).

Rastaworld existed along a spectrum of tourism types in Ghana, including the bubble of leisure tourism, which included "infrastructural arrangements that permit the professional reception of guests—such as hotels, lodges, personnel, logistics— plus those arrangements making the travel of tourists possible; travel agencies in the sending as well as the host countries, transport facilities and a massive internet information business" This "tourist bubble," "protects the visitor from the unfortunate aspects of a destination while permitting some view to the outside" (Van Beek & Schmidt 2012:23). At first glance, Rastaworld would seem the opposite of leisure tourism, defined as it was by seeming insertion into Ghanaian life in search of cultural difference. In this, it more closely approximated cultural tourism (Graburn 1989, MacCannell 1976, Smith 1989, Urry 1990) and functioned as a touristic borderzone. Defined by Bruner (1996) as "a zone of interaction between natives, tourists and ethnographers which is described as a creative space that allows for the invention of culture" (1996:157–179), such zones by necessity contain cultural brokers—informal intermediaries between the tourist bubble and the local community. Rastas were such intermediaries.

Yet like leisure and cultural tourism, Rastaworld had its own temporal rhythms and activities, which were starkly different from the everyday realities of working Ghanaians. Indeed, the foreigners who sought out Rastas rejected the Ghanaian tourism bubble in their search for authenticity and difference, but they also rejected the strictures of everyday Ghanaian life. Their notions also rest upon a simulacrum—for travelers sought a hyper realism, or magical realism, that felt more immediate and momentous than the quotidian everyday they had left behind. The Rasta version of Ghana was an example of the spaces to which modern mobile subjects travel precisely in search of that which has been lost, for if mobility and change are central to modernity, then the imagined tradition of the immobile "other" in both space and time represents stability (Adey 2010, Salazar 2010). To find this desired, longed-for feeling, tourists sought out the simulacrum of "pseudo-events," geographically located within but radically removed from the everyday realities of the places visited (Boorstin 1964, see Baudrillard 1994).

Because Rasta cultural practice drew so heavily on diasporic resources, its relation to heritage or roots tourism, the "homecoming" of diasporans to the homeland from which they were severed by the cross-Atlantic slave trade, was complex

(Bruner 1996, Hartman 2008, Holsey 2008, Reed 2014). The return of diasporans to Ghana had begun at independence, when President Kwame Nkrumah had invited diasporan intellectuals, leaders, and activists to help build the nascent state and to take part in his project of national *cum* Pan-African *cum* diasporic emancipation (Gaines 2006, Taylor 2019). This sense of Ghana as the site of "homecoming" to Africa was reformulated by President Rawlings' 1990s regime in neoliberal form. Like Nkrumah, Rawlings viewed the African diaspora as essential to Ghana's economic development in the neoliberal economy. Marketing Ghana to diasporans as the "Gateway to Africa," the Rawlings government sought economic investment from the diaspora. It organized a tourism system that centered pilgrimage and commemoration, turning the 32 slave forts (or "castles") that line Ghana's coastline into heritage tourism sites, especially those at Cape Coast and Elmina (Bruner 1996, Ebron 2002:Chapter 7, Hartman 2008, Holsey 2008, Reed 2014). It was here that Ghanaian playwright Efua Sutherland inaugurated the first Pan-African Historical Theatre Festival (PANAFEST) in 1992.[12] As the anecdote that opens this chapter illustrates, Rastas often performed within the heritage tourism system.

On the other side of the spectrum were volunteerism, the practice of visiting Ghana through a meaningful "help" framework, and music tourism, "travelling to listen … as opposed to traveling to look" (Stokes 1999:141, Ebron 2002). Volunteerism's primary goal was growth or transformation of the self through encounter with the other. Here, the consumption of cultural difference was filtered through an "authentic, hands-on experience," imagined as experiencing rather than seeing, as being inside rather than outside (Swan 2012:242–244, see Sin 2009, Wearing & Ponting 2009). Music tourism was similarly based on experience, in this case learning to drum (usually done by male visitors) or dance (usually done by female visitors). Many aborofo who occupied Rastaworld had also, or would also, come to Ghana as volunteers or students of drumming and dancing, and they shared a desire for experience from within rather than seeing from without. Touristic borderzones throughout the world are defined by such collisions of imagination (Adey 2010, Salazar 2010).

Tourist art (the objectification of mental images tourists bring to the places they tour) and romance tourism (sexual or emotional encounters with a person who embodies this image) are frequently found in these in-between—transitional, transcultural, transactional—contact zones, for tourists need an object of desire or a subject of a quest (Ebron 2002, Frohlick 2007, 2013, Jacobs 2009, Kempadoo 1999, Meiu 2008, 2017, Pruitt & LaFont 1995). Both depend on an imagined authenticity that can be found in zones somehow untouched by Western influence (Schildkrout 1998:10, see Beidelman 1997:8, Phillips 1994:42, Steiner 1994:115). The ongoing circulation of images in such borderzones renders them available for translation and transformation, radically expanding the materials available to shape lived experience.

Ethnographic Encounters with Ghanaian Artworlds

Throughout this book, I discuss the parallelism that defined Rasta and touristic self-productions in the contact zone of Rastaworld. Similar parallels existed between Rasta and ethnographer, and between ethnographer and tourist, producing a hall of mirrors effect that generated numerous insights while I was conducting my field-work in Ghanaian artworlds. Faye Ginsburg calls this the "parallax effect," noting that "if harnessed analytically, slightly different angles of vision can offer a fuller comprehension of the complexity of the social phenomenon we call culture and those media representations that self-consciously engage with it" (Ginsburg 1995: 65). Moving through different contexts during my fieldwork in Accra, shifting from one performance of self to another, witnessed by others as I did so, respond-ing to their responses, I often experienced a sense of "mimetic vertigo" (Himpele 2002:302), as all of us—Rastas, other artists, tourists, myself as ethnographer, and others—performed ourselves in the intercultural arenas of Accra. This straddling of worlds and shifting of selves by subjects possessed of different types of mobility would hence become a central theme of my analysis.

My fieldwork in Rastaworld began in 2001, although I did not know it at the time. Rastas were the first Ghanaians, outside of airport personnel, to directly ad-dress me the first time I landed in the country in June of that year. I was in Ghana for the first time on a two-month summer visit as the soundperson on a documen-tary film crew. I stayed on to conduct preliminary research, with a view toward developing a dissertation project on Ghanaian arts. I had met Moses while I was on the film crew. A young drummer raised in Accra, he took on the informal labor typical of Rastas by becoming our cultural intermediary, acting as fixer, translator, and guide whenever we had to shoot scenes in the capital. Moses had grown up inside the Accra Arts Centre, so when I told him I was interested in the arts, he insisted on taking me there. Rastas were the first to greet us when we entered. As a young, foreign, female, first-time visitor, it made sense that I encountered Rastas as Ghana's forward-reaching, outstretched hand.

Following this introduction, I returned in 2003 to stay in Accra for 15 months to conduct ethnographic research. My plan was to work at the Arts Centre, which I viewed (with relief) as a bounded (literally, by walls and a coastline) microcosm of the larger field of Ghanaian artistic production, its physical spaces seeming to materi-alize that field's disjunctures and overlaps (Bourdieu & Johnson 1993). But it would prove impossible to stay within these supposed boundaries. The walls had holes. As a young, White woman, my ongoing presence at the Arts Centre confounded people and challenged me. I was not a tourist passing through to shop for souvenirs, nor was I a dealer in African arts and crafts. I was not a friend or girlfriend to a Rasta, yet I kept returning ("You're back!?" was a frequent early greeting) and disappearing again. Partly due to the discomfort of being a White woman permanently ensconced in that space, my initial research project gradually expanded beyond the local tourist craft trade. What began as an ethnography of crafts at the Arts Centre grew to in-clude the fine artworld, then to include gender relations between oboroni women and Ghanaian men, and finally to focus on Rastas as the ultimate embodiments of many of the forces I saw operating throughout the different scenes. Therefore, I come to this project as an anthropologist of art—not of music, performance, tourism, race, or gender—and whatever insights I have come from that viewpoint.

Thus, when Moses first welcomed me in 2001, I entered a complex field in which I embodied a principal organizing figure: the oboroni woman. Like many of the tourists, expatriates, curators, and art buyers, I was a gendered bearer of foreign possibility. Like them, I was an outsider learning about Ghanaian culture, using art as an entry point. And like them, I was an object of intense perusal, made deeply, constantly aware of my female Whiteness, my fundamental bodily differ-ence, and adjusting my presentation of self within different contexts. In this sense, I was like Rastas, engaged in an ongoing project of interactional, intersubjective, intercultural self-making, doubly aware of my body as both subject and object, as simultaneously the skin I lived in and the screen upon which others projected their imaginings.

Chapter Summary

Like Rastahood itself, *Aesthetic Practices in African Tourism* is a practice of crafting in the round. In a parallelism of technique, it borrows from different theoretical literatures to think through and understand Rastahood. I move the reader around the Rasta figure, each chapter describing a facet, so that by the end, the figure as a whole can be seen anew. Like Rastas, I construct and deconstruct how subjects make their lives through making objects and performances under late capitalism.

Chapters 1 and 2 take place at the Greater Accra Regional Centre for National Culture, locally known as the Accra Arts Centre, or Arts Centre. Chapter 1, "Geography Is Destiny: Craft in Accra," explores the multiple histories that were materialized in the space of the Accra Arts Centre to become equally available for Rastas to draw upon. The Centre is in the oldest, most densely layered part of the capital, and the chapter moves upward through layers of time to discuss this part of the city under the Ga (the region's original inhabitants), as part of the British colony of the Gold Coast, through Ghanaian independence and the creation of national culture, and as it trans-formed into a tourist art hub in the neoliberal 1980s and 90s. Each of these historical moments left a trace in the palimpsest of the site and became a resource for Rastas to draw upon. The Arts Centre emerges as a site where "culture" has repeatedly been objectified and recirculated, making it an optimal training ground for Rastahood. Chapter 2, "Men at Work: Craftwork, Masculinity, and Precarity," explores the script of craftsmanship and adult masculinity Rastas inherited but were unable to fulfill. Through the words and experiences of elder craftsmen, specifically Ewe carvers and Hausa dealers, I explore the transmission of knowledge and the socialization of youth into men through apprenticeship, and then shift to the pressures placed upon this sys-tem under neoliberalism, which made it increasingly difficult for carvers to make a living and support their apprentices and for youth to survive in Accra as apprentices, causing displaced youth to turn to hustling, a practice of hovering in public spaces to connect potential clients with specific shops for a small dash. Elder and younger men registered such changes as a deep loss that extended far beyond shifting labor practices to a foreclosure on the known future.

Having set the stage for understanding how and why Rastahood developed, the next three chapters present, deconstruct, and analyze the Rasta figure. Chapter 3, "From Elephants to Drums and Beyond: Object, Performance, Mobility," takes up where Chapter 2 left off. The inherited labor trajectory toward adulthood now cut off, apprentices-turned-hustlers find a way forward in the arrival of a new tourist art object, a non-Ghanaian West African drum called the jembe that had been popularized in the West through diasporic acts and the spread of world music. The drum transformed age-based relations of knowledge transmission and social reproduction and signaled shifts in the labor market, materializing the entrepreneurial temporality of the new era, and the precarity of social relations and identities. To better connect with young tourists, foreigners, Ghanaian youth introduced performance into their art form, having learned the skill from senior drummers at the Arts Centre.

Chapter 4, "Styling the Rasta Self," and Chapter 5, "The Affective Labor of Crafting Freedom," shift from production and formal performance to subcultural style and affective labor as additions to the artistic arsenal. In Rastahood, the stylized body emerged as a locus of agency and cultural expression. Rasta style allows youth to look both ways, simultaneously critiquing their position within Ghanaian society and solving the problem through the new oboroni-Rasta relations the style enabled. Chapter 4 focuses on the former, exploring how reggae and Rasta style emerge in 1990s Accra as means to articulate a new generational identity. Beyond craft production and performance, Rastas engage in immaterial labor, including work as guides, translators, mediators, and fixers for the tourists with whom they connected. Chapter 5 examines how Rastas curated the tourist experience and then moves beyond this to explore the reciprocal, if unequal, relationships between Rasta men and oboroni women through Rasta and oboroni definitions of freedom. In this chapter, the work of Rastahood comes to fruition, enabling youth to connect directly with aborokyire. The book ends with a leap forward into the present, out of Ghana, and beyond Rastahood, as men, now in middle age, look back upon the path that Rastahood created.

Notes

1 I use the term "neotraditional" to refer to the repurposing and restaging of pre-existing cultural forms for new audiences in new, usually broader contexts.

2 This book is based on ethnographic research conducted in Ghana, the United Kingdom, and Denmark between 2001 and 2016 as well as ongoing digital communication with its participants. I first went to Ghana in 2001 to conduct preliminary research and work on a documentary film. I lived in Accra between 2003 and 2005, where I worked with various communities within the tourist and high artworlds. I also conducted research in Cape Coast and Kokrobitey. I returned to Ghana for shorter stints in 2006, 2009, and 2012. International fieldwork took place in the United Kingdom in 2005, 2006, 2009, 2010, 2012, 2014, and 2016, in the United States in 2006, and in Denmark in 2012.

3 I coined the term "Rastaworld" while conducting research to explain my feeling of entering and exiting a world that was not a place. I later discovered John Jackson's *Harlemworld* (2001). Jackson samples this hip hop term to address Harlem's symbolic power in the popular imagination and the ways, as a place, Harlem extends "beyond its strict geographic boundaries" (Jackson 2001:10). The term is also in dialogue with Jacqueline Nassy Brown's "racialized geographies of the imagination" (1998:291). See Shalini Shankar's *Desi Land* (2008) for another "world" that transcends geography through media and popular culture.

4 Following Mikhail Bakhtin's definition, the aesthetic function is defined by a situated-ness outside experience and lived reality (Bakhtin et al. 1990). This situatedness outside is crucial—the aesthetic always responds to and completes reality. The art object, "a thing obtrusively demarcated in time and space from all other things" (Bakhtin et al. 1990:275), subjects an already cognized reality "to concrete, intuitive unification, in-dividuation, concretization, isolation, and consummation, i.e., to a process of compre-hensive artistic forming by means of a particular material" (1990:281). For Bakhtin, the aesthetic object is the instantiation of activity on the part of the creator. It thus highlights the performative, iterative, embodied nature of aesthetic action. "In form I find myself, find my own productive axiologically form-giving activity, I feel intensely my own movement that is creating the object" (1990:309). Art objects are, "in a phrase of Robert Goldwater's, primary documents; not illustrations of conceptions already in force, but conceptions themselves that seek–or for which people seek–a meaningful place in a repertoire of other documents, equally primary" (Geertz 1983:99).

5 The term Northerner encapsulates an array of groups, some within the borders of Ghana and some to its North. These include Dagomba/Dagbani, Tamale, Dagarti, Frafra/Fare Fare, Moshi/Mossi, Kusasi, Grunshi, Sisala, Lobi, Wala, Gonja, Konkamba, Busanga/Mosanga, and Bosu people from The Upper West, Upper East, and Northern regions of Ghana and from neighboring Burkina Faso as well as Hausa people (Holsey 2008:96, see Quayson 2014). The Hausa are a Sahelian people chiefly located in the West African regions of northern Nigeria and southeastern Niger, but diasporic communities exist throughout West Africa along centuries-old trade routes. In Accra, the term Hausa was often used to demarcate a diverse group of Muslim migrants, including Yoruba, Hausa, and others from the country's Northern region and the borderlands it shares with Burkina Faso. The Ewe are an ethnic group from the Volta Region in Eastern Ghana (as well as being the largest ethnic group in neighboring Togo).

6 As Bosiwah et al. (2015) note, "The Akan word 'aborofo' (Europeans) came into existence in Akan through a derivational process. However, two schools of thought exist concerning the actual meaning of the word 'aborofo'. Those who perceive the colonial masters to be wicked (due to their activities in the country) interpret the word as aborɔ-fo(ɔ) 'wicked people'. The other school of thought has it that the Europeans got the name aborofo because they came to the Gold Coast by sea, hence, a-borɔ-fo(ɔ) (i.e. from behind the horizon). The word àbòrofó could either come from the base àbóró 'wickedness' or the root bòrɔ 'behind the horizon'" (2015:1). These authors lean toward the latter. See Holsey (2008) and Pierre (2012) for discussions.

7 Along with the authors cited in text, my reading of this localized Whiteness is informed by work the scholarship on race and gender in colonial contexts, especially that of Anne McClintock (1995), Ann Stoler (1995), and, in the case of Ghana, Carina Ray (2015).

8 bell hooks's conceptualization "eating the other" is relevant to this idea of Africa. For hooks, "eating the other" entails the consumption of difference in controllable, commodified doses, which allows majoritarian subjects to spice up their lives and accrue cultural capital through temporary play in alternative playgrounds without relinquishing their mainstream position-ality and the power that attends it (hooks 1992:366–367, see Hall 1993: 105).

9 I build this term on Roland Barthes' "Italianicity," the specifically non-native image the French have of Italy, which is produced through a redundancy of signs (Barthes & Heath 1978). My use differs sharply from Africanicity as an oppositional ontology within Afri-can postcolonial studies as well as Africanness, a sense of the African aspects of oneself in Africa and the African diaspora (Dei 2012, Mafeje 2000, Epprecht & Sigamoney 2013, Spronk 2009). In my use, as in Barthes', it is an image made by an outsider.

10 With the exception of one biracial, North American woman, all the women I met who were involved with Rastas were White.

11 See Dor (2014), Reed (2016), Salazar (2010), Steiner (1994), Sunkett (1993) for rel-evant studies on world music, tourist art, and tourism. See Salazar and Jayaram (2016) for studies of mobility and Salazar and Graburn (2014) for tourism imaginaries.

12 Funded by the Ghanaian government and the African Union, PANAFEST was the first international festival to be held in Africa since FESTAC (World Festival of Black Arts and Culture) had brought the "Global African Family" and Lagos, Nigeria in 1977 (Ofori-Ansa 2003). According to the festival's website, PANAFEST was likewise intended to re-unite "the African Family" through performing arts, speeches, and ritual events commemorating the slave trade. PANAFEST was privatized in 1997, but it continued to be held every couple of years. In 1998, President Rawlings instituted Emancipation Day on August 1, making Ghana the first African country to commemorate the end of chattel slavery in British colonies in 1834. Subsequently, it was funded by the Ministry of Tourism. In 1999, PANAFEST and Emancipation Day were celebrated together at the height of the summer tourist season to enable the largest number of diasporic Africans to participate (Reed 2014:ch. 3). Holding such Unification and Emancipation festivals and PANAFEST indeed brought a windfall of foreign currency to Ghana, but it also resulted in a clash of visions between local and foreign interests. For analyses of the complexity of such homecoming politics, see Hartman (2008), Holsey (2008), and Reed (2014).

1 Geography Is Destiny

Craft in Accra

Rastas like those in this book have been ubiquitous worldwide ever since the world music boom of the 1980s. Any tourist who has taken a tropical beach vacation in the Caribbean, Central or South America, or coastal Africa within the past 40 years has likely encountered a dreadlocked young man looking to connect with tourists by offering them a way into an authentic experience of "local" culture. Rastas are not all the same, however. Their localness (and that of their Rastahood) matters because personal survival and social reproduction are historically and geographically specific problems. Along these lines, Rastahood attracted Ghanaian youth for Ghanaian reasons, with these youth drawing upon aspects of Ghanaian culture alongside diasporic and global resources available in the country.

In Ghana, the 1990s ushered in a period of economic, political, and public liberalization. Its effects were most obvious in the cosmopolitan city of Accra, the capital port city through which many people, ideas, and goods have flowed in and out of the country over the centuries. Liberalization specifically introduced new opportunities and conundrums for male youth coming of age, who sought to resolve the problem of becoming and being a man without family money or higher education to fall back on. Rastahood represented one of the solutions, along with varied other possibilities mapped onto the bodies of young men, who created overlapping stylistic, sonic, and somatic maps of the city as they moved through it (Feld 1988). And so, by the turn of the millennium, clean-shaven, shorn-haired, conservatively dressed Pentecostal youth (Meyer 2004, Meyer 1999) were crossing paths with hiplifers in low-slung, baggy jeans and shades (Osumare 2012, Shipley 2013) and dreadlocked Rastas in tie-dyed Bob Marley t-shirts and beads. These youths lifted their prayers, hymns, sermons, raps, and reggae songs up and over the others around them to engage the possibilities of a life within this neoliberal context.[1]

For their part, Rastas created a Rastaworld, as I am calling it, which expanded and contracted across a range of sites in Accra. But its core was the Greater Accra Regional Centre for National Culture, locally called the Arts Centre, a sort of material holding place for many of the ideologies and histories of art and craft in Ghana and a microcosm of the city's multiple, at times conflicting spatial logics. Founded at independence in 1957, the Arts Centre was the largest, best known, and most visited of Ghana's ten regional cultural centers, all of which were sites of simultaneous cultural nationalism and cultural tourism. As an extension of tourist art production emanating

DOI: 10.4324/9780429244568-2

from the Centre, Rastahood depended on a self-conscious self-representation through certain objects, performances, or embodiments, which functioned synecdochically as representations of Africanicity for foreign consumption. To this end, culture had to be objectifiable, uptakeable, recirculable, dividable, with isolated pieces excised from their context of use, simultaneously retaining the aura of past alliances and accruing power from new remixes. The Arts Centre was the ideal place for the development of Rastahood as a mode of cultural production, because the Centre had already been the site of such objectification under multiple politicocultural regimes and in various artistic realms; hence it still contained remnants of these imaginaries for Rastas to draw upon by the time they entered it in the 1990s. Culture everywhere is invented (Hobsbaum and Ranger 1983), but how this happens is always a local matter, and the Arts Centre's "multi-synchronicity" (Jameson 1991) was definitive.

The Arts Centre sits on Accra's coastal southern boundary, in the heart of historic Accra, surrounded by national monuments. Its main gate opens onto John Evans Atta Mills High Street, the city's oldest main thoroughfare, near major markets and ministries. Its footprint has changed over time, but in the late 1990s and early 2000s, the sprawling, walled complex contained a *souk*like handicraft market, a large textile market, a kiln, and a colonial-era building housing administrative offices, and several performance spaces. A post office, an internet café, and a forex bureau abutted a mahogany-shaded plaza, where one could find itinerant hawkers and food sellers. A sandy lane lining along the Centre's northern boundary was lined with open-faced workshops. Known as Carvers' Lane, it constituted the Producer Section of the Centre.

By the mid-1990s, the Arts Centre was Ghana's largest craft market, offering tourists an overwhelming array of Ghanaian and West African goods, including cloth, jewelry, beads, woodcarvings, stools, masks, brasswork, baskets, leatherwork, calabashes, and drums. It thus presented a kind of materialized, temporally flattened (partial) history of West African arts and crafts. To walk the Arts Centre complex (as Rastas would do every day) was to walk through a series of spatially co-extensive pasts, each rendered available in the present, time equalized by space in an ontology of ever-changing, accruing pastness (Alcock 2002).[2] The Centre was the densest, most layered point of the artistic city, a palimpsest of all its histories. It is no wonder, then, that the Arts Centre is where Rastas came of age and were exposed to every aspect of the Rasta assemblage except Rastafarianism itself. It is where Ghanaian youth learned about drumming and dancing and managing performance troupes, drummaking and managing carving workshops, performing, producing, and selling "culture," and managing relationships with foreigners.

If one arrived at the Arts Centre as a potential customer on any given day in the first decade of the new millennium, one would surely have been greeted by a Rasta. Sitting in traffic on John Evans Atta Mills High Street, waiting to enter the Centre through its white, concrete main gate, one might first notice a young Rasta weaver stationed on a wooden bench along the exterior wall of the Centre, working bright, synthetic thread of red, green, black, and white, into tank tops, wristbands, and belts. Then, among the hawkers approaching the car in standstill traffic, offering everything from car phone chargers to toilet paper, a dreadlocked youth might push a blackened wooden carving of a market woman in profile against the car window. Finally, entering the gate either by car or on foot (having given up and exited the taxi), one would instantly be greeted (some tourists might say accosted) by young men, hands reaching out to be shaken, saying, "Hello my friend, how are you?" These Rastas would attempt to ascertain what one was looking for so that they could bring said item to person or person to item. Overwhelmed by the attention, one might glance around to see more young men sitting under the trees, surrounded by goblet-shaped drums in various stages of production. Some of the drums would be haphazardly strewn about, while others would be arranged in neat lines or piles. One would gradually discern the repetitive sounds of the young men's work: the rhythmic scratch-scratch of wooden drum shells being sanded, the knock-knock of wood on metal on wood, the slight creaking of wet goatskins being stretched, or ropes tightened, and the whisper of wood dust sifting down onto the sand. One might take this layer of people, activities, and objects as the whole, yet it belied the multiplicity of craft histories that brought Rastahood into being. Moreover, the young men who addressed one so similarly and were so uniformly attired and employed arrived at this place through very different paths. Let us consider three such paths below.

Paths to the Arts Centre and Rastahood

Moses: A Child Is Born at the Arts Centre

When Moses first heard I was in Ghana because I was interested in studying the arts, he got really excited. "'Accenta!' You must come to 'Accenta'! I live there!" This was what I misheard him say. Thanks to his animated enthusiasm, I became fascinated by the place even before I understood that—unfamiliar with his accent—I had misheard him saying "Arts Centre." Then I went through a second stage of incomprehension, as I puzzled over how someone could live at an arts institution. It turned out that not only did he currently reside there, but he had also been born and raised within its walls.

The origin of the Arts Centre and how Moses Danquah came to live there date back to Ghanaian independence in 1957. At the time, its first Prime Minister, Kwame Nkrumah, immediately took over the European Club, also known as the British Officers' Leisure Club. He intended to use it as a base for a national Arts Council; in time, it became the Arts Centre. This was not the first such center in Ghana. The Asante Cultural Centre was already established in the city of Kumasi at the end of the colonial period, in 1955, under the direction of anthropologist Alexander Atta Yaw Kyerematen, music teacher, musicologist, and composer Ephraim Amu, and choreographer, dancer, artist, and educator Albert Mawere Opoku. It included a museum of material culture built according to traditional Asante architectural design and supported an Asante dance group and other cultural activities (Hess 2003, Johnson 1979). The Asante Centre was lauded for awakening interest in traditional arts among the educated classes throughout the regional capitals of the country. Before then, their notions of cultural activities had mainly focused on European "ballroom dancing and a few school concerts and dramatic performances founded on little or nothing that is indigenous to Ghana" (Antubam 1963, Collins 2001). However, Nkrumah's government viewed the Asante Centre as a possible manifestation of Asante's claims to independent nationhood, making its status problematic. So Nkrumah undertook complex negotiations to bring the Asante Centre into the national fold as just one of ten regional centers, focusing resources into developing the Arts Council complex in Accra (Antubam 1963). This included moving Public Works Department (PWD) workers and their families into the barracks that originally housed British officers behind the Leisure Club, to be known as the Annex.

As Moses told me, his father, Danquah, was one of the government workers who was granted family housing in the PWD Annex. By 1980, the year Moses was born, an entire community was residing within the complex. Some of the PWD residents worked outside the Centre, while others ran businesses inside it. Two chop bars (restaurants), two spots (bars), various seamstresses, a tailor, a barber, a hair salon, and other small businesses served the residents, craftspeople, and traders who themselves increasingly rented shops within the Arts Centre. Over time, those businesses served to bring together Arts Centre employees, residents, and visitors and, in so doing, became important sites of exchange.

For his part, in addition to working as a driver for government officials, Danquah was a talented drummer from an Ewe musical family. With other government workers, he formed All Clans, an ethnically mixed drumming and dancing troupe that included Ewe, Ga, Asante, and Dagbon dances. All Clans rehearsed at the Arts Centre along with other drumming and dancing ensembles (Dor 2014, Schauert 2015). Rehearsals took place after work in the main theater and on weekday evenings in the inner plaza, with different groups alternating use of the spaces. Potential clients would drop by the rehearsals to check out the various troupes and then hire them to play at state-run events or private occasions, such as funerals, weddings, or outdoorings (naming ceremonies for infants).

Danquah's children thus grew up surrounded by music. Moses remembers being taught drumming and dancing, "before [we] could walk," leading him and all his siblings to become drummers or dancers. They dispersed among the troupes at the Arts Centre, often switching from one to another as needed. Moses himself, from a very young age, displayed prodigious talent and an obsession with drumming. When I first met Moses, he was not yet a Rasta, but a short-haired, 21-year-old who favored crisp white t-shirts, a shy youth who was already performing as lead drummer in ensembles composed of older men. In this, he was atypical.

Most of the Arts Centre resident children entered ensembles by proximity, starting out as child acrobats and slowly gaining skill as dancers and drummers. They attended school all morning, worked in the afternoon carrying out small tasks for adults, including craftspeople, and rehearsed in the evening. For their afternoon work, they earned small "dashes" (or tips), which went toward paying their school fees and expenses. From their musical-residential life, they learned the contours of the various roles (i.e., director, drummer, dancer, acrobat) and of various tasks entailed in being a member of a troupe. Kids observed the daily lives of people involved in ensembles, including their interpersonal dynamics and social hierarchies. They understood what it would take to run rehearsals, book performances, and forge relationships with clients.

The potential for travel, to places in concentric circles expanding outward from the Arts Centre, was also part of the life and appeal of these ensembles. For Moses, this sort of opportunity came by joining Black Star, one of the first groups founded by youth. Unlike the senior men who performed at large-scale, private, theatrical, or state functions, most of the audience for young musicians was foreign. They played for customers at the Arts Centre or tourists in coastal towns and cities and then leveraged the relationships they developed into foreign tours. This allowed Moses to first leave the country to go on tour to Europe when he was 22 years old. Playing abroad was exciting and expanded his experience, but it did not provide much economic stability; leaving Moses to return from Europe again and again to the Arts Centre, to rejoin the daily struggle of making ends meet. At home in Ghana, he spent the next decade in this rhythm of coming and going.

When I spent time with Moses, I always ended up visiting parts of the Arts Centre I had never seen before, as though the Centre was infinitely divisible. He often hung out at the usual Rasta haunts: certain drum shops on Carvers' Lane and Akwaaba Restaurant, the Rasta-dominated spot between the Main Gate and the top of the Lane. But he spent even more time in the back areas of the Centre which tourists never saw. He was the one who brought me to the Annex, to spots and chop bars where people whom I never saw elsewhere in the Centre ate and drank, and to a building where children in a drumming and dancing group rehearsed. He was the one who introduced me to a seamstress from whom he had commissioned Rasta-themed clothing to sell abroad, to the barber who used to shave his head before he began to (dread)lock his hair, and to others who had grown up with him in the Annex. Thus, the Arts Centre remained a touchstone for Moses. Long after he had moved elsewhere, he would come back to the old neighborhood and to all the neighbors who were owed greetings.

Samson: A Hawker Strolls In

While Moses was, in his own words, "born into it," Samson wandered in. Kofi Donkor, known as Samson at the Arts Centre, was born in Accra in 1979, the eldest of three children. His mother, an Accra-born Ga, raised the children in two rented rooms in a house in a compound, while his father, an Akan, lived in Togo throughout his childhood. Like many working-class women, Samson's mother started out as a petty trader, eventually improving her situation by selling vegetables at the market. This introduced Samson to the trade. As a small boy, Samson helped his mother by selling pineapples on the street. Since such hawking was usually done by girls, people were moved to see a boy doing it. "People would buy from me," he said, "because they wanted to help a small boy helping his mother."

Samson's mother always told him if he should ever need anything or run into any trouble, he should go to the Arts Centre, where she had a friend who ran a chop bar frequented by performers. While trouble did not prompt him, he would do the rounds of the Arts Centre community, as part of his regular hawker's path through town. And one day, the teenage Samson walked by his mother's friend's chop bar and saw a boy about his own age sitting on a bench, making a goblet-shaped drum. Samson was engrossed by the boy's skills. That night, he went home and told his mother he wanted to make drums.

This was 1999. Samson was part of a generation of working-class Accra boys who were out of school, their families unable to afford university, and unclear on their prospects for the future. Working in the informal economy seemed an obvious possibility, but the best option, selling foodstuffs, was less viable for men than for women.[3] The Ghanaian informal economy was segregated by sex, so some occupations with higher rates of pay are closed off or heavily stigmatized for men because they are considered domestic, feminine industries rather than masculine ones (Akorsu 2013:165–166). Women's work includes trading goods of all kinds, including foodstuffs (Azu 1974, Clark 1994, Kilson 1974, Oppong 1974, Robertson 1984). Thus, whenever I pointed out the overwhelming maleness of craftworkers, Samson and others would comment, "Women can sell."

"Me-myself, I have done work men don't do," he commented once as we discussed the topic of work with two other Rastas. "Have you ever seen a man selling *fufu* [a beloved local food made of boiled, pounded cassava and plantain]? I did it."

The goblet-shaped drum opened a different world of possibility for Samson. He immediately befriended the boy, named Shem, and began to learn drummaking from him. Of equal importance, the chop bar was owned by Shem's mother and happened to be located around the corner from the Annex where Moses lived. Samson recalled:

> That is when I met Moses and everybody, because all the different groups in Arts Centre—when they come to rehearse, people pass through the house, they always pass through the house to maybe Shem's mother, or maybe Shem's sister, who is also a dancer. And Moses's house was around the corner, so we were all living there, sometimes we go and fetch water there, you know? Go and eat. Go and sit. Like children. It was like—compound—like how you live in a community. The people who live there know each other.

Others soon joined Samson and a young PWD resident named Gideon, both of whom had recently begun working with Shem. They were brothers, cousins, friends, youth on the outs—all looking for work. Many of these young men had been born in Accra to migrants from elsewhere in Ghana, making them multilingual and of dual ancestry, but steeped in Ga language and culture. All of them spoke at least two languages—Ga, the language of the Accra ethnic group, and Twi, the Akan language of the Asante, which serves as Southern Ghana's lingua franca. Many also spoke a third home language, usually Ewe or Hausa, and most taught themselves English. Some, who were still in school, began building drums in the afternoons and using the money to pay for their school fees. Others, who had graduated or could not afford to complete school, were "sitting in the house doing nothing," at the beck and call of their elders. As Samson explained, "When you are growing up at Arts Centre, everyone can tell you to do things: 'fetch this,' 'do that,' 'call him.' And you do it, and it is only later that you realize you have learned a skill. At first it is just something you do." All these boys, looking for something to do, found each other at Shem's mother's chop bar. Roughly ten of them would end up forming the core group of Arts Centre Rastas.

For the future Rastas, there was no formalized path to their emerging profession. Unlike other craftwork in Ghana, there were no workshops, no master-apprentice relationships, through which the knowledge of jembe drummaking could be transferred. The boys taught themselves and learned together. At first, they worked on single drums, earning just enough profit from selling one drum to buy the raw materials for the next, improving on their skills a little bit with each attempt. At the same time, they began to make a place for themselves, taking over unused areas of the complex, including making drums under the trees behind the Arts Centre. And they fostered a market by befriending young travelers interested in drumming.

The timing worked out well. Jembe drums were gaining popularity with the rise of world music during the 1990s, which introduced the drum to Western audiences. As the drum became increasingly popular, so did the demand for them. The young drummakers went from producing 10 drums, to 50, then 200 drums per contract.

In their work, the young men benefited from the presence of other master craftsmen and craftshops at the Arts Centre. Although jembe production was not formally practiced there, nor did it follow traditional modes of transmission and training, the craftshop model, like that of the drumming and dancing ensemble, offered a materialization of what adulthood, masculinity, and independence could be. Everyone wanted a shop of their own.

Samson was part of the collective that opened the first youth-owned, youth-operated drum shop at the Arts Centre known as Black Star and formed the eponymous drumming and dancing group in which Moses performed. The two entities would eventually fission into numerous jembe shops and drumming ensembles, but Black Star, as a shop, remained most successful. Samson was an excellent businessman, and his drums were of high quality. Over time, he taught himself to play and teach drumming. He also had a knack for intercultural translation, effectively mediating Ghanaian and aborofo needs, desires, and ways of doing things. Hence the Black Star shop became a central stop-off point for young travelers in Ghana, dreadlocked oboroni boys learning to drum and braided oboroni girls learning to dance. They came to learn music and buy drums to take home with them, but above all to experience what Rastaworld had to offer in the way of Ghanaian culture. For this reason, the shop was frequently the last place travelers hit, hoping to be seen off with a performance before heading to the airport, allowing their departure to mark the significance of their trip. Thus, you never knew when impromptu drumming lessons and jam sessions would burst out of the Black Star shop.

If walking with Moses meant moving inward, ever deeper into Arts Centre, walking with Samson turned it into an ever-expanding space. He was always getting calls on his phone, always had multiple projects in the air. He moved between situations with a focused, fast-but-slow energy. We would stop at one of the two shops he rented in Carvers' Lane, where youth were always at work filling large orders for drums or loading container trucks bound for the port. We might then go to the welder to pick up parts or visit the purveyor of skins, both of whom had shops in the back of the Centre. Then we would go to town to run an errand, taking a back route out of the Centre to Akuma Village, the Rasta village-restaurant-nightclub-performance venue where Samson lived. Finally, we would end up at a rehearsal for Black Star, his drumming and dancing group, or at a Reggae dance event. In this way, Samson became a guide, introduced me to Ga Accra, Ga elders, elder drummers, and "real" Rastas.

Noah: The Long Bus Ride South

Unlike Samson, Noah Akarma started out a stranger to Accra. Noah was born in Bolgatanga, the capital and craft center of the Upper East Region, known for its leatherwork, straw weaving, and hand-loomed men's smocks, all of which were popular throughout Ghana and throughout the craft export trade. Noah was the second of five children, three of whom were girls. Although he excelled in school and aspired to a career in science, his parents could not afford to pay for his education. So he put himself through junior secondary school, first by helping his mother sell staples and snacks at a little roadside stall, then by hawking batteries and other small goods on foot, having purchased them with a small loan from his mother. Later, when his father fell ill and could no longer support the family, he chose to drop out. His senior brother could have supported Noah, but then his sisters would have been unable to go to school. "I love my sisters very much, and if they did not go to school, maybe they would be with men, and do stupid things, or be influenced." This is how he ended up in Accra, at least in part to earn money to keep his sisters in school. As a member of the FraFra ethnic group, Noah likely followed old labor migration routes paved by his forebears as they too sought work in the southern capital, port of opportunity (Hart 1973).

On arriving in Accra, many migrants from the North ended up at the Arts Centre. As a national center for fine arts, its administrators welcomed craftsmen and traders from all Ghanaian regions as early as 1979. According to numerous accounts by Arts Centre craftsmen and administrators, by 1989, woodcarving, basketweaving, and rattan weaving were all occurring on-site in Carvers' Lane and handicrafts and textiles were being sold in two large, open-air markets. Each of these trades was loosely controlled by a particular ethnic group: Ewes carved wood; Northerners from Bolgatanga made baskets; Akan, especially Asante people, sold textiles; and Northerners sold handicrafts. Noah had no particular attraction to Culture as constituted in these crafts; for him as for many others, craftwork was simply the most realistic way to earn money. Knowing that senior members of his family were already involved in straw weaving and had a shop at the Arts Centre, he took the long bus ride to Accra in 2000 to join them. When he arrived, he discovered that the shop was in crisis and the senior men who ran it were barely making ends meet.

Out of necessity, Noah turned to two new jobs in the neoliberal Arts Centre economy: hustling and drum piecework. Each represented forms of survival work for those with few options. Hustling, for its part, was a direct result of changes to the Arts Centre and broader Ghanaian economies. By 1999, the national cultural centers had begun charging rent to craftsmen and dealers and increased the number of shop permits so the centers could maximize returns and stay afloat. At the same time, shopowners in all sectors felt the sharp bite of intensified competition, as more and more people made or sold the same thing. The economic pressure created by these two forces meant that shopowners had to find new ways to bring customers to their goods. Tourists in turn entered the Arts Centre, sometimes searching for specific craft items, sometimes not knowing where to go, but always confronted with a bewildering number of shops selling similar things. Enter the hustler, who would connect with tourists and shepherd them to particular shops, where he would earn a dash (a tip, or percentage) for a completed sale. The other survival job available was piecework. When a drum shop got a large contract, it outsourced the basic aspects of production (sanding, roping, shaving, and pulling) to temporary workers who were paid by the piece.

Hustlers and drum pieceworkers were some of the most precarious workers at the bottom of the Arts Centre labor pool. Described by older shopowners as having "no place" and "no craft," they were despised embodiments of the current state of affairs and were sometimes viewed as criminals. Despite this, Noah and his schoolmates who had moved south were able to stake a place for themselves in the Arts Centre. They usually sat on an old mahogany log that lay in front of Akwaaba Restaurant. From this strategic vantage point, they could watch for business to appear from all directions, with the entrances to the Arts Centre building, the Handicraft Market, the Textile Market, and Carvers' Lane all in view. The young hustlers would spend most of their time sitting, watching, waiting, and hoping for customers to come or something to happen. In the process, they formed a tight group, supporting each other and sharing what meager money they earned, so they could eat at the end of the day. Eventually, one of them, usually with the financial help of an oboroni girlfriend, might be able to set up a little jembe shop of his own in the smaller half of a former carver's subdivided shop at the top of the Lane. But each small rectangle of independence was precarious, coming and going, with little ability for the former hustler to control his fate.

As for Noah, in the first years I knew him, when he worked at the Arts Centre, he was in a sort of stasis, an embodiment of waithood. We never walked around together. We always stood in place, first at the log in front of Akwaaba, and then in front of his tiny shop nearby. The geography we discussed was far from where we stood. It included the North, which Northerners nostalgically described as the way Ghana used to or ought to be, and aborokyire, as a place where a future might be possible.

***

Moses, Samson, Noah: A walk out the front door, a circuitous path around the city, and a bus ride traversing the country from north to south brought three teenage boys to the Arts Centre and all the possibilities for and constraints on social and economic improvement that it offered. But their story is not only a contemporary one. It rests on layers of history that generated the conditions of possibility in which they lived. The Arts Centre contained many of these layers; it was like a matrix in which not only artifacts but also practices, histories, and ideologies existed as traces (Alcock 2002:2, Glissant 2002). It was a gathering place—"Think only of what it means to go back to a place you know, finding it full of memories and expectations, old things and new things, the familiar and the strange, and much more besides" (Casey 1996:24). It was defined by its "cityness" (Simone 2009:3) that carried the potential of transformation through encounter for both the space and its inhabitants (Massey 2005:116).

For Rastas, this place mattered because it made these traces available as resources, in the sense that Brown (1998) defines them, only here these traversed depth in time rather than breadth in space. In this way, the Arts Centre distinguished Rastas who came of age on its premises, enabling them to create their complex figure, which, despite the approbation it generated, was a logical permutation of what the Centre consistently produced: tourist art objects made through the reinterpretation and recombination of existing elements, which, though not "authentic," resonated with recognizability.

Excavating a History of the Accra Arts Centre

Depending on whom I sat or walked with and where we walked in the Arts Centre, many elements of 500 years of Ghanaian history would rise to the surface in people's stories.[4] As Ato Quayson states in his own urban history of Accra,

> To understand the city as a whole we are obliged to keep shifting our analysis to encompass the moving pieces of a geographic jigsaw puzzle, which, depending on which way we turn, divulges slightly different perspectives on urban space. This method of interpreting Accra is much like what Chinua Achebe recommends in *Arrow of God*: if the world is like a mask dancing, you cannot see it by standing still (1964, 46).
>
> (Quayson 2014:67)

It was likely a combination of my presence and quiet moments that spurred accounts of earlier times, but these reflections on the past also depended on what object I held, what story it carried, and who claimed its authorship. A painting of two abstracted, elongated market women, hanging in the main building in the Arts Centre, might prompt an academically trained artist or administrator to tell of the 1960s, when this was one of only two galleries in all of Accra. A mask in the handicraft market would provoke a Hausa dealer to retell how Hausas had brought the handicraft trade to Accra. Walking under the mahogany trees in the central plaza where hustlers idled, waiting for a customer, would call forth an elder carver's memory of his own time under the tree, when carvers were the first people tourists encountered upon entering the Arts Centre, and the trade was healthy.

Some people recalled even earlier times. One day, while I sat in Samson's second shop, a thin, dapper, old man arrived in search of a cane. It was a blazingly hot afternoon, and the man was tired, so Ezekiel (Samson's cousin who was overseeing the shop while Samson was in Europe) dusted off a workbench and invited him to sit down. Ezekiel bought a water sachet from a young girl walking by and gave it to the old man before setting off to look for a cane for him in the handicraft market. After greeting each other, the man and I sat in silence, squinting out at the glare. Then he said quietly to me, "You know, this place used to be called Polo Grounds. It was only white people here, and if a black boy came, he would be sent away." And before and during after this, Ga ceremonies had been held on the land where he and I now sat, and people remembered that as well, for the Arts Centre sat near the "the most ritually saturated part of the Ga landscape" (Nat Nunoo-Amarteifio in Quayson 2014:85).

Layer 1: Ga Accra

The Arts Centre sits on the oldest, most historically layered, culturally saturated part of Accra, Ghana's capital since 1870 and the country's "most important site of exchange, transaction and power," as historian John Parker notes in his study of the Ga territory that became Ghana's capital (Parker 2000:xvii). "Throughout its history, Parker continues, Accra has been a quintessential 'middleman' state, its inhabitants mediating a variety of transactions across geographical, political, and cultural frontiers" (Parker 2000:xviii). Thus, in Accra, the African "urban *problematique*" of "who shapes the city, in what image, by what means, and against what resistance" (Mabogunje 1990:122–123) is most deeply stratified. As a result, outsiders who come to Ghana seeking an Africa that is "pristine, untainted and essentially rural" (Parker 2000:xxiv) can be bewildered, finding instead Accra's fundamental admixture—its ongoing blend of flows, influences, and borrowings.[5] They find what the Ga epistemology itself presents: the outward orientation of the city and the sea, civilization as fundamentally grounded in processes of exchange and constitutive of a cultural order which must be defended (Parker 2000:6).[6] For Rastas, this perspective is intimately familiar.

Hence, intercultural mediation and exchange, so central to Rastahood, have been present on the Accra coast from a very early stage. The earliest known settlers of the coastline were Kpeshi, about whom little is known. But they were in turn joined by "very mixed" peoples who migrated to the coastal plain from the east at different points in time (Dakubu 1972:100) who came to be known as Ga.[7] No definite time of first arrival has been identified, but it appears that waves of Ga arrived from the 13th century and were well established by the 17th, having six settlement townships or *man* (pl. *manjii*) along the coastline that included Ga Mashie, Osu, La, Teshie, Nungua, and Tema (Odotei 1991:61, Quayson 2014:38).

Ga traders of the 1980s repeatedly told anthropologist Claire Robertson that "there is no such thing as a pure Ga" (1984:28). Oral tradition and historical sources confirm their assertion. As the region grew, this organically developing Ga inclusivity came to be embedded in the social structures that shaped Accra's built environment.[8] Hence, for example, the township of Ga Mashie, the heart of Ga Accra, included seven autonomous Ga *akutsei* (quarters), each of which maintained a distinct identity, and three of which had been founded by foreigners (Odotei 1991:63, Pogucki 1954, Quayson 2014, Robertson 1984:43–50).[9] Thus, strangers from other ethnic groups were incorporated into the Ga polity and engaged in Ga cultural practices, including life cycle rites and annual festivals. The Ga further expanded this admixture of people by developing relations (partly through intermarriage) with other African groups that were drawn to the coastal trading zone. Thus, "as an ethnic group the Ga are of mixed origin" (Odotei 1991:61), containing the practices of intermarriage and incorporation of foreign elements that would later be central to Rastahood.

Geography also plays a role in how this polity and culture of admixture emerged. Ga territory sits in a niche between the forest and the sea, bordered to the south by the Gulf of Guinea and to the north by the Akwapem escarpment 15 miles away, on an arid coastal plain that lacked large-scale agricultural possibilities or tradable natural resources such as ivory or gold. Its strength lay in the location between land and sea.[10] But with the arrival of the Portuguese in 1482, the resource-inhibited location of the Ga allowed them to enter the Atlantic economy as traders (Parker 2000:xxvi).

Quickly noting the Ga "aptitude for trade" (Robertson 1984:29), waves of incoming Europeans turned the area into a major trade center, and Ga operated as intermediaries between the maritime European powers and inland communities, especially Akwamu, Akyem, and Ashanti traders (Quayson 2014:38). Economic activity was significant enough for the Europeans that they established three trade lodges in the westernmost towns along the coast: the Dutch built Fort Crevecoeur (Ussher Fort) at Kinka in 1649; Denmark-Norway constructed Fort Christiansborg at Osu in 1652; and the British built Fort James in Jamestown/Nleshi in 1673. These forts stimulated the growth of Accra as the Ga commercial center (Akyeampong 2002:41, Odotei 1991:62, Robertson 1984:27). Until 1680, the Ga capital was located at Ayawaso, 11 miles inland from the coast, and the Abonse Market was a central exchange for European, coastal and inland goods. But beginning in the 1640s, the Gas sought to monopolize trade with Europeans, leading to conflicts with their neighbors and culminating in the destruction of Ayawaso and Gas' migration to the coast (Odotei 1991:62, Quarcoopome 1993:20, Quayson 2014:39).[11] Subsequently, as the Europeans increased their control over local affairs and opened immigration to non-Ga peoples such as the Fante, Ewe, and Akwamu, each of the settlements around the forts—British Accra (Jamestown/Nleshi), Dutch Accra (Ussher Town/Kinka), and Swedish-then-Danish Accra (Christiansborg/Osu)—developed distinct identities and contributed to Ga interethnic admixture (Quarcoopome 1993:20). By the mid-18th century, the Ga were playing a central role in moving enslaved people into the Atlantic slave trade. Over 1,209,000 people left the Gold Coast until Britain's Abolition of the Slave Trade Act of 1807 (www.slavevoyages.org). The arrival of groups such as Yorubas and the Tabon—Afro-Brazilian former slaves who landed in Accra in the 1820s and 1830s—brought urban skills and transnational networks to Ga society (Quayson 2014:44). When Basel missionaries arrived in 1828, they began training Accra men in various trades and teaching arts and crafts in mission schools, which resulted in a pool of people with primary or secondary education who were also at least somewhat artistically trained. Teaching trades and crafts was considered as important for missionization as preaching (Herppich 2016), and the Basel missionary community was particularly responsible for bringing this sensibility to the region.

As a result, by 1870, Accra had become the main source of skilled labor along the coast, with Ga workers trained in metalsmithing carpentry, masonry, architecture, coopering, and shoemaking. These new skills again opened the Ga to a larger world, pushing them to travel as far as Congo for work and bring back new commodities, which would become key elements of the economy (notably the cocoa pod, brought by Ga blacksmith Tetteh Quarshie in 1876 from the island of Fernando Po) (Robertson 1984:30). Meanwhile the Ga retained the advantage of Accra being the only real city on the Gold Coast (Konadu-Agyemang 2001) and its land still belonged to the Ga state or to specific Ga lineages, an advantage that would continue until land became alienable under colonial rule (Sackeyfio 2012:298).

For Rastas born or raised in the capital, like Moses, Samson, and others in this book, Accra's Ga identity was central even if they were not born Ga. They spoke Ga among themselves, and they lived and worked in the oldest, most Ga-saturated part of the city. As people regularly told me, the eastern side of the Arts Centre, where the Producer Section sat, had been Ga holy land and farmland until as late as 1979, and many recalled Ga customary rites being performed where their ancestors had been buried (see Parker 2000). Beyond these connections, though, Rastas manifested Ga traits that had shaped Accra for centuries: an epistemology of intercultural mediation and exchange, an identity based on admixture, and a mental geography where the continent met the sea and what came from beyond the horizon.

Layer 2: British Accra

The recombinatory model of culture that undergirds Rastahood resonates not just with Ga attitudes. The next historical layer in the Arts Centre's palimpsest was colonial (specifically British), and it was crucial to Rastahood in several ways. In the mid-19th century, Britain consolidated its presence by buying out the Danes in 1850 and the Dutch in 1872, declaring the Gold Coast Crown Colony in the name of the Company of Merchants, then subsequently in the name of the Crown, in 1874. The disparate territories that comprised the British Gold Coast thus united formerly separate political, ethnic, and cultural groups from which Rastahood drew both its community and its forms, as new migrants from elsewhere in the colony came to Accra with their regional artisanal skills.[12]

From the late 17th until the late 19th century, Britain governed through direct rule, providing a small local elite access to European culture and civilization in exchange for allyship in the colonial enterprise. Members of this coastal elite "served as teachers, clergy, doctors, civil servants, law clerks, journalists, and academics within the growing European imperial project in West Africa," and their missionary education allowed them to be not only intellectuals but also merchants and legislators (2013:22, see Fyfe 1992, Gocking 2000). But as Frederick Cooper has observed, the British could never control the civilizing mission, which produced not only allies but also critics (Cooper 2002:Chapter 1).

However, the consolidation of British power in the late 19th century led to a transition from direct to indirect rule. As theorized by Mahmood Mamdani, while the former's intention was to "shape the world of the elite amongst the conquered population," the latter aimed "to shape the world of the colonized masses" (Mamdani 1999:865). In so doing, indirect rule ousted the local elite, which direct rule and the British civilizing mission had created, and the threat it posed to colonial rule. Coining the term "hegemony on a shoestring," Sara Berry explains how the shift to indirect rule was caused by financial stringency on the part of the British Exchequer, which required colonial officials to limit the number of British personnel hired (Berry 1993:24). Berry notes, "One obvious way to cut costs was to use Africans both as employees and as local agents of colonial rule. African clerks and chiefs were cheaper than European personnel; also, by integrating existing local authorities and social systems into the structure of colonial government, officials hoped to minimize the disruptive effects of colonial rule" (Berry 1993:25).

Indirect rule, in which British officials sought to use "tradition" as the basis for their administration, depended upon "neatly bounded, mutually exclusive, stable cultural and political systems" (1993:27) of an imagined African past, as opposed to the reality of dynamic communities defined by porous boundaries engaged in multiple conflicts over power and resources (Akyeampong 2000). This shift was pragmatic, only later justified as a coherent ideology (Beidelman 1982, 2012). By the end of World War I, British official had consolidated this view of an Africa organized into neat, contained, tribal categories, culminating in Lord Lugard's *The Dual Mandate in British Tropical Africa* (1923), which laid out this system and the new imperial mission with which to govern it (1993:28). But, as Frederick Cooper has noted, this mode of governance only exacerbated the volatility of African societies: "Trying to confine Africans to tribal cages … created the very sort of danger administrators feared" (Cooper 2002:20). Atlantic port cities like Accra were always in flux, as people moved and forged new identities in urban contexts. According to Mahmoud Mamdani, indirect rule masked a racial dualism "in a politically enforced ethnic pluralism" (Mamdani 1996:7), but this was never a simply horizontal pluralism. A supplementary difference was made between people considered indigenous to an area (for instance Gas in Accra) from "native foreigners" or "aliens" (Northerners in Accra) and granting land rights only to the former, regardless of how long the latter had resided in said area (O'Rourke 2012, Quayson 2014:52). Thus, while indirect rule's "tribal cages" could not repress the relational identity formation happening in cities, creating and incorporating tribal categories into colonial state structure had dire effects on postcolonial identity and belonging.

Rastahood's appeal to young aborofo would depend and draw upon the constructed racial/cultural difference between African and European, while the development of the Accra craft trade more broadly would be shaped by the rights and modes of employment of foreigners in Accra.

Indirect rule was enacted through economic and political policies, education, and urban planning. In 1877, the British transferred their administrative headquarters from Cape Coast to Accra, uniting British, Dutch, and Danish Accra into a single town (Quarcoopome 1993:21).[13] British life centered around the High Street, where one found a large judicial complex erected alongside commercial trading houses, residences, the British Bank of West Africa, an Anglican Church, and the Basel Mission (Hess 2000:37–38). But most other urban arrangements were haphazard, and Europeans considered the city unsanitary. These problems were compounded as increasing numbers of European administrators and merchants arrived and were joined by their wives. In response, the colonial government took over the city's development, essentially disenfranchising the local townspeople.[14]

The British established a new urban logic that spatially represented the division between European and Native, who previously had lived among one another: racial segregation done in the name of "sanitation" (Parker 2000, Pierre 2012, Quayson

2014), and expressing what Ato Quayson describes aptly as "the essential coupling of the civilizing impulse with that of nausea for local life" (Quayson 2014:41).[15] The European town and administration area (the European Central Business District) now overlay the Ga city described above, oriented around the harbor and rendering Jamestown a commercial warehouse district (Pierre 2012:27, Quayson 2014:73). High Street itself became purely commercial (Hess 2000:40), and a newly renovated Christiansborg Castle was turned into the governor's residence and government offices (Quarcoopome 1993:24).

"Health segregation" measures justified the development of the open land between Kinka and Osu into an exclusively European quarter in 1898 (Pellow 2001:62). First called "New Site" and then Victoriaborg, this quarter would "mark the explicit European preference for residential racial segregation" (Quayson 2014:80). To the north of the High Street, it contained administrative, business, recreational, and residential buildings built and arranged in a recognizably British tropical style (Hess 2000:39, Jackson and Oppong 2014, Parker 2000:97, Sackeyfio 2012:298; see Amoah 1964, Berkoh 1974, Brand 1972). Homes were built on large plots of land, included a separate servant quarter, and had water, electricity, and a network of roads to the ministries (government offices), banks (High Street), and the central business district. The coastal expanse south of High Street was left relatively undeveloped, making it a spacious site of colonial leisure. It contained leafy parks, the European Club building (which had the only built stage in Accra), Polo Grounds, and a rifle range (Quayson 2014:80). This site of colonial leisure and racial exclusion would be transformed into the Greater Accra Regional Centre for National Culture.

Indirect rule shaped the future Arts Centre and Rastahood through its influence not only on urban design but also on education. After World War I, the Gold Coast's cacao industry produced astounding profits, making the colony an economic success story.[16] Drawing on the colony's reserves, Gold Coast Governor Sir Gordon Guggisberg (1919–1927) instituted a ten-year development plan to include major infrastructure projects. One of the most important was Achimota College, which opened in 1925, 7 miles to the north of the city. By 1928, the prestigious Achimota College had expanded to include primary, secondary, and university-level classes and a teacher training course (Steiner-Khamsi and Quist 2000:277). The college was a direct response to British fears of losing control over educated elites as well as the populations of colonial Africa's growing cities, perceived by the British as sites of new social and political problems arising from "detribalization," politicization, and over-education. A new type of education became a top priority, and Guggisberg's "greatest interest" (Wallbank 1935:234).

Achimota's imagined, ideal student was neither the average African who would return to the village to farm in the off-season, nor was it the elite, urbane future leader of the people. Instead, it was the "chief's son" (Coe 2002:26), the "educated fieldhand" (King 1971:49), and "safe artisans and rural teachers, unchallenging to the white status quo" (King 1969:660).[17] These "betwixt and between" figures would be trained in the city but return to the country to mediate the dualisms of colonial rule. Having learned a "respect for science and a capacity for systematic thought," these colonially engineered "agents of modernization and providers of stability" (Coe 2002:26) were nonetheless expected to remain "African in sympathy and desire for preserving and developing what is deserving of respect in tribal life and custom, rule and law" (Foster 1965, quoted in Harrod 1989:148), "some elements of the old ... grafted on to the new" (Wallbank 1935). By grafting the best of European education onto the best of African culture, Achimota would "train Gold Coast boys and girls to be leaders of their people" (Slater 1930:463), trained "so that instead of flocking into the towns they may go back to their villages, as chiefs, teachers, housewives, farmers, medical assistants, and artisans" (in Steiner-Khamsi and Quist 2000:280). Achimota would enable the design of a partial African modernity in line with British goals.

In reality, the pedagogy at Achimota was influenced by several, at times conflicting, fields of intellectual production, including British colonial ideologies of indirect rule, White American philanthropic and missionary thought regarding Negro education, the demands of a growing educated, urbanized African elite, and British modern artists' utopian notions of art, craft, and culture. The result was an "adapted industrial curriculum" (Steiner-Khamsi and Quist 2000).[18]

"Culture" was a key element in the paradoxical attempt to shape colonial African subjectivities.[19] The ideology linking art to rule that developed at Achimota during this period, and the art education infrastructure that resulted, would continue to be instrumental in shaping the Arts Centre and Rastahood. All the later cultural institutions which define Ghanaian art worlds came into being at the college. It was the first place where "art" was taught and instructors trained to teach it. Its Teachers' Training College (eventually moved to the city of Winneba) was the institution through which many of Ghana's first artists passed, for "back then," artist Ablade Glover explained, "you could not make money from art, you had to teach." Achimota was also the place where British teachers influenced by the British Arts and Crafts Movement reconceived precolonial arts as "craft" (cottage industries set up on-site at Achimota were early experiments to transform craft into modern, small-scale "industry" and are precursors to today's government-run Arts Centres). Achimota's fundamentally British curriculum was "Africanized" by also including instruction in the languages, drumming and dancing traditions, and arts and crafts of four Ghanaian ethnic groups—the Ewe, Asante, Fante, and Ga—which Achimota administrators viewed as intrinsic to African selfhood. But the inability of British teachers to teach these subjects, their essentialist rhetoric, Achimota's refusal to hire African teachers (Jenkins 1994:183), and its (mostly unsuccessful) outsourcing of instruction to local artists (Coe 2005, Shipley 2015) resulted in a fundamental othering of what they imagined to be "African." Unsurprisingly, Achimota's elite students resisted this aspect of the curriculum.

Art was first formally taught at Achimota College in the 1920s. Achimota's early art tutors were heavily influenced by the European Arts and Crafts Movement. They combined the movement's suspicion of modernization and industrialization with racialized, colonial logics of detribalization, their vision filtered through preservationist panic and nostalgia. In this ambivalent rendering, the arts curriculum was imagined as a breakwater between the Africans and modernity. It would enable students to stay "in close contact with African life" so that, after completing their schooling, they would return to the villages to work as chiefs, teachers, farmers, and artisans (Pickard-Cambridge 1940:148). Following a jigsaw theory of culture (Beidelman 1982), which assumes "that it is possible to manipulate pieces, removing them at will without disturbing other sections and to replace them by new sections" (Foster 1965:165, quoted in Coe 2014:55), Achimota administrators sought to convert the traditional arts into crafts, "without the old high religious sanctions but with what is almost as high, the tradition and seal for fine craftsmanship and honest labor" (Stevens 1930:154). At the same time, they were aware that Achimota was producing the only academically trained artists in the country (Stevens 1930:156, 1935).

In 1936, the artist H.V. Meyerowitz, along with his artist wife, Eva (an ethnographer and sculptor), left South Africa and moved to the Gold Coast (Antubam 1963, Kwami 2013). Meyerowitz would "revolutionize art thought and the approach to art and crafts teaching in the country" by converting Achimota's secondary school art program and teacher training college into a "proper School of Art and Crafts, offering a three-year course in specialist art and crafts teaching with a bias in Ghanaian, African art traditions" (Antubam 1963:199). The "Meyerowitz Approach" (Kwami 2013:223) shared Stevens' interest in "convincing art students that they have an African viewpoint to express and to share with the rest of the world" (2013:223).[20]

The colonial logic of African difference and antimodernity, and its view of the arts (Culture) as the salient African contribution to the rest of the world, would prove remarkably resilient, appearing in millennial tourist narratives of Rastas, whom tourists perceived as more authentically African than the majority of the city's residents.

Layer 3: Northerners' Accra

In addition to the Ga and the British, Northerners, often collectively called Hausa, greatly influenced Accra and the crafts trade at the Arts Centre. The Hausa are originally Sahelian people, chiefly located in the West African regions of northern Nigeria and southeastern Niger. A large global Hausa diaspora also exists, including many communities found along centuries-old trade routes in West Africa. However, in urban Ghana, the term "Hausa" is applied by non-Muslims to a wide variety of groups who migrated south to live in *zongos*, the bustling "strangers' quarters" that exist for Muslim migrants in every large Ghanaian town, who wear the same fashions, and who use Hausa as a *lingua franca* (Pellow 1985, 2001). At the same time, the inclusive Hausa identity construct of urban Ghana cannot be assumed when examining other contexts (Adamu 1978, Cohen 1969, Pellow 2002, 2003, Schildkrout 1978).

Thus, for example, Muslim traders, likely of Mande origin, were already present on the coast when Portuguese merchants first arrived in 1482, and during the slave trade, enslaved Northerners had cultivated land and worked as porters. After abolition, Northerners were hired in palm oil cultivation (Ntewesu 2005). By the mid-18th century, the number of Muslims on the coast expanded through three population movements, which included Hausa who had migrated south to Kumasi following the kola trade. Salaga Market, established within Ga Mashie in 1874 for Northern migrant vendors (of kola and other goods), points to their centrality within the city's life (Quayson 2014:66, 76). By the 1880s, over 1,500 Yoruba and Hausa Muslims lived in Central Accra, mostly in houses on Zongo Lane, Okanshie, and Horse Road. They provided cheap labor as porters, construction workers, servants, and members of the army and the police.[21]

By 1907, the Central Accra Muslim collectivity, now supplemented by Kanuri, Fulani, Nupe, and Wangara peoples, presented as a "socio-religious as well as spatial unity" (Pellow 2001:427). They formed a large and seemingly unassimilable group by intermarrying, dressing alike, living according to Muslim law, knowing little of the local language, and refusing to war against one another. As persons of different ethnic origin, Northern immigrants were categorized as "strangers" under colonial Ga customary law; they were also beholden to the Ga, through whom Northerners received custodial rights to land they occupied (Pellow 1985:430–431).[22]

This Northerner community, and its stranger status, would eventually become important to the history of the Arts Centre as well as to local understandings of tourist art. Several of the Arts Centre elders I interviewed, including both woodcarvers and handicraft dealers, stated that at some point during the colonial period, and certainly by the 1940s, the Muslim community in Zongo Lane, collectively known as Hausa, had introduced a handicraft trade to Central Accra that would become the basis of the Arts Centre craftworld. As they explained, because they did not own land, men of this community were well situated to enter the trade: they often worked for Europeans as watchmen or security guards, thereby becoming acquainted with the European taste for collecting "primitivist" mementos such as masks, stools, and other ritual objects. Trade in these objects would have been anathema to coastal animists and Christian converts, who believed such objects retained dangerous powers. Muslims, however, understood themselves as immune to such spiritual forces, and the objects' only power lay in their objectification of European imaginaries of Africa, a desired Africanicity that could be commodified. In this way, Northerners found an advantageous sociocultural position in the colonial order, between other Africans and Europeans, and from this position, they learned to see both the animist object and the European mental image—and translate one into the other.

Engaged in this manner in the craft trade, dealers began commissioning new genres of objects from traditional Ewe carvers already working in Accra. They drew from a network of Ewe commercial carvers from the northern Volta Region, who, by the 1950s, had moved to the coast to settle in the port cities of Takoradi and Accra, following a long-used trade route from the Ewe heartland to the north (Lawrance 2007). Having completed their apprenticeships in their home villages, newly minted Ewe carvers would join their masters in individual workshops dispersed throughout Accra, where they carved all kinds of objects out of hardwoods local to the Volta, including mahogany, rosewood, ebony, and teak. Aside from carving functional objects and their own traditional ritual forms, they would sometimes copy masks, or photographs of masks found in postcards or artbooks, brought to them by Hausa traders from as far away as the Congo, Gabon, and Sierra Leone (Wolff 2004:123). Hausa traders also trained Hausa youth to "finish" the masks by sanding, staining, polishing, accessorizing, and artificially aging them to make

them look authentic, made in and for a context untainted by Western influence
(Schildkrout 1998:10, see Beidelman 1997:8, Phillips 1994:42, Steiner 1994:115).
They would finally take the finished goods and "go bungalow," that is, hawk them
in the European residential area, to which they were granted access due to their
employment as domestic servants, while other Africans were prohibited from entering (Pierre 2012:30). Zongo Lane therefore became a point of conjuncture and
production for a particular assemblage of creativity, travel, economy, and desire.
The Arts Centre would become a similar nexus for carvers who made such objects,
and later for the Rastas who opened jembe shops there.

Throughout the transition from colonialism to independence, the itinerant handicraft trade continued to grow. I learned about the next step in the trade's institutionalization through a possibly apocryphal story told to me by both carvers and
traders. In 1957, Queen Elizabeth sent her sister Princess Margaret to represent
England in the Ghanaian independence ceremonies. A few years later, in 1961,
Queen Elizabeth herself visited Ghana for the first time to consolidate its allegiance
to the Commonwealth. The story goes that Elizabeth—or perhaps Margaret—encountered handicraft dealers and was so impressed by their work that she organized and brought them to the High Street, where they set up small kiosks along
the roadside in front of the Ghana Commercial Bank. Thus relocated, they would
be seen by European passersby. The Ewe carver Godwin Ametewe explained to
me that "any time an expatriate needs something, they come to them. So, that is
the place where the *real* business start from—High Street." A few hundred meters
down the road from the site of Nkrumah's Arts Council and later Arts Centre (see
next section), the Hausa handicraft trade may have remained outside the officially
sanctioned space of the state, but it was only *just* outside. The contact zone of Central Accra had accrued another layer. Hence, the path traveled to Accra by these
earlier icons of Africanicity—the product of so many different people's labors and
imaginations—mirrored that which the jembe drum, the primary artifact of Rastahood, would subsequently take when it arrived on the same land.

Layer 4: Nkrumah Era

On March 6, 1957, Ghana became the first sub-Saharan nation to gain political
independence, transforming, from Britain's "model colony," into a symbol of in-
dependence, anti-imperialism, and Pan-Africanism for the entire Black world.[23]
The first official public utterance of this historical shift is a speech Prime Minister
Kwame Nkrumah made at midnight that day, standing at a podium on the Old Polo
Grounds in Victoriaborg, adjacent to Arts Centre land.

> At long last, the battle has ended! And thus, Ghana, your beloved country is
> free forever!... From now on, today, we must change our attitudes and our
> minds. We must realise that from now on we are no longer a colonial but free
> and independent people.... We are going to demonstrate to the world, to the
> other nations, that we are prepared to lay our foundation—our own African
> personality. ... You are to stand firm behind us so that we can prove to the

world that when the African is given a chance, he can show the world that he is somebody! … We have awakened. We will not sleep anymore. Today, from now on, there is a new African in the world!

As Prime Minister of tropical Africa's first independent state, Nkrumah faced the enormous task of constituting a national identity that would unify Ghana's diverse populations, represent the Pan-African aspirations of the continent and the diaspora, and prove to the Western world that Africa could rule itself. He would pursue this goal, in part, by using the country's relative stability and wealth (Ghana was the world's leading cocoa producer at the time) as a springboard to industrial development, planning massive infrastructure projects that would move the country away from a single cash crop economy and constitute it as a Pan-African socialist state.[24]

Alongside such ambitious political and economic policies, Nkrumah also planned to harness "culture" into what architectural historian, writer, and former mayor of Accra, Nat Nunoo-Amarteifio, described as a "total war" against colonialism:

> Kwame Nkrumah … came from a tradition which is no longer around, and this was very classically educated people. … And he believed that for a nation to progress you needed not only scientific advancement but also cultural advancement. Kwame Nkrumah also came from a generation that was produced just before the Second World War. Now the Second World War was a war in which nations literally fought with everything they had, from shovels to guns to Beethoven. I mean, it was a total war … And his generation grew up in that intellectual atmosphere, where the whole nation was engaged in that struggle for survival. It's not surprising that when he came home after the war and started the drive for independence, he did not focus only on a political aspect but he also drew into his fold the culture of the nation. There was a generation which went into war with him that threw in their genius into helping him achieve independence. And there were a lot of artists involved in this group.

As has been noted by many, what Nkrumah wanted was to render the new nation-state more visible to its citizens and the world (Schauert 2015), and to "make a particular contribution to the totality of culture and Civilization" (Nkrumah 1958).

Nkrumah's conception of culture, as indicated by Amarteifio, was generational, shaped within and against colonial logics. Like much of Africa's ruling national elites in the early years of independence, Kwame Nkrumah had been educated in colonial schools and subsequently at universities in America and Europe. At Achimota College, he had learned that every national culture should be based on native folk traditions (Botwe-Asamoah 2005, Sherwood 1996), which should in turn be recontextualized to rid them of potential ethnic divisiveness and transform them into a "usable past" (Fuller 2014). Nkrumah thus envisioned a modern African art incorporating diverse ethnic genres into European media to serve both anticolonial and pronationalist aims. Combined with symbols of modern statecraft and European royalty, such art would create a unitary culture—a new "African personality" (Nkrumah

1961, 1997). The new identity was emblazoned upon Ghana's architecture, flag, currency, stamps, and other objects of statehood. It was visualized in murals, paintings, and monuments and performed through the spectacles of music, dance, and theater.[25]

In the years leading up to independence, Nkrumah was already focused on the use of arts to create a national identity. He wanted to "show that Africans had a culture worth emulating," according to Amarteifio. Hence, in 1954, Nkrumah set up the Ghana Arts Council Interim Committee to oversee the cultural events surrounding Independence, where not only Ghanaians but also many foreign dignitaries and visitors would witness what the future of Ghana, and by extension, Africa, would mean. In 1958, after independence, the Interim Committee became the Arts Council of Ghana, with the now permanent aim of encouraging the development of a national culture.

From that first moment on the Old Polo Grounds on the day of independence itself, Nkrumah made Central Accra a critical part of the nationalist arts project. Sociologist, ethnomusicologist, and member of the original Arts Council, Joseph Hanson Kwabena Nketia recalls that "the first thing that we did when we became an Arts Council was to take the European Club, which is the present Arts Centre. It used to be the European Club for the expatriates, so after Independence you could take it over." The European Club was especially useful because "it had a stage, it had everything, you see. And we did not have a theatre anyplace. Then we could have a place where people could come and see things. So that was beginning of the arts movement ...."

In seeking to create a new culture for the modern state which would nonetheless be "closely linked to our traditions and express the aspirations of our people at this crucial stage in our history" (Nkrumah 1961), Nkrumah remained focused on leading artist-intellectuals of the period to design the nation's symbolic images and found its key institutions. Achimota-educated artist Kofi Antubam designed the new state's visual culture, including two Presidential Seats of State, the State Sword and Mace, reliefs at the National Assembly, and murals at the newly built Ambassador Hotel, where visiting dignitaries were housed during the independence celebrations (Kwami 2013). The National Theatre Movement was created by playwright Efua Sutherland. And again, in concert with the overarching effort to create a new national narrative, this movement focused on the performing arts to "reflect the traditional culture of Ghana and yet develop it into a living force firmly rooted in and acceptable to the new generation in Ghana" (Antubam 1963:210, Shipley 2015).

Other important national arts and education institutions were established soon after independence. Composer-musician Philip Gbeho, who composed the national anthem, became the founding director of the Ghana National Symphony Orchestra in 1959. Scholar-dancer Albert Mawere Opoku became artistic director of the Ghana National Dance Ensemble in 1962. Nketia had already been made head of the School of Music, Dance, and Drama at the Institute of African Studies at the newly autonomous University of Ghana at Legon in 1961. And that same year, Kwame Nkrumah University of Science and Technology was established in Kumasi. But as a remnant of colonial planning, the arts programs remained split between the two universities, with visual arts in Kumasi and performing arts at Legon (Allman 2013, Schauert 2015, Shipley 2015).

The arts were not divided up in the built space of the Arts Council, however. On that single proscenium stage in Accra, Nkrumah's cadre of radicals created new forms of theater, music, dance, and acrobatics and provided school children and other amateur performers with guidance and rehearsal space (Coe 2005, Dor 2014, Schauert 2015, Shipley 2015). Meanwhile, fine arts, mostly paintings, were exhibited in a gallery at the Arts Council compound and in a room in the Ambassador Hotel, which, according to senior artists Ablade Glover and Kofi Dawson, were the only two exhibition spaces in Accra at the time. Eventually, the Arts Council (a.k.a. Arts Centre) would become a critical site for superseding the exclusionary colonial culture of entertainment with an anticolonial culture capable of dismantling colonialist images of Africa. However, according to Amarteifio, the transmutation of traditional arts into signs of modernity and nationhood was slow to develop beyond "an elite, a small group, endeavor."[26] Nketia concurred:

> There was a lot of interest [among] the literate—you know, in those that have left the traditional culture—interest in going back to the traditions, but also in using the traditions in new forms....But this was just among the intellectuals and literate people, and it was not then a mass movement, you see....We knew that the traditional things were very important, but we were also very interested in what *we* would do with our experience. So the transformation idea was also implicit in the creative thing that emerged. And in fact Nkrumah showed interest in *that* even more so than in the preservation of traditional cultures.

Drumming and dancing, however, were two aspects of the developing national culture that, even from the beginning, extended beyond the intelligentsia. Numerous amateur groups began to train at the Arts Council under Gbeho's supervision. One in particular, the Asiedu Ketete Cultural Troupe, brought together Ga, Ewe, Akan, and Northern musicians to perform a variety of rhythms from throughout the new Ghanaian territory, formalizing the intraethnic exchange called for by the political climate of indigenous fusion. Master drummer, dancer, and teacher Kobla Ladzekpo remembers:

> The whole political climate then was an encouragement for us to go back to our indigenous traditions. 'Sankofa' [Go back and retrieve it]. It was a very dynamic and very encouraging time period for us young people. So we picked up the challenge and learned as much as you can from our other neighbors, tribes [ethnic groups]. Now we had formed all kinds of groups at the Arts Council.
>
> (Ladzekpo 2007, quoted in Dor 2014:241)

These groups would play key roles in the later development of Rastaworld.

Handicrafts were not included in the original designs for the Arts Council. The national cultural elite avoided indigenous craftwork because of the colonial association with primitivism. Crafts were sometimes incorporated into national or regional celebrations and spectacles but were represented at the Arts Centre only on a very small scale, starting with individual demonstrations by a carver, kente weaver, and jeweler. Nonetheless, eventually, the artisans at these small demonstration units were allowed to produce crafts for sale and later, according to then Associate Director Samuel Ashong, the Arts Centre provided them with materials and salary to collect the profits from their sales.

By the mid-1960s, the optimism of independence had waned. Nkrumah's regime had grown increasingly autocratic, banning regional political parties and suppressing the cocoa farmers upon whose production Ghana depended. The export agricultural economy was in collapse and public debt increased ten-fold. This led to a military coup d'etat, on February 28, 1966, led by the National Liberation Council (NLC), whereby Kwame Nkrumah was overthrown.

The following 15 years were defined by chronic regime change and increasing economic instability.[27] Throughout this chaotic period, the cultural institutions that had materialized independent Ghana's Pan-Africanist national project fell into disrepair, as each regime was either indifferent toward or actively disavowed Nkrumah's legacy. The Arts Centre became a quiet place. Arts Centre craftspeople later recalled that people farmed, priests administered sacred rites, and criminals conducted business in the undeveloped land east of the Arts Centre buildings.

Layer 5: 1980s Rawlings Populism

On December 31, 1981, Flight Lieutenant J. J. Rawlings led the Armed Forces Revolutionary Council (AFRC) in a second coup against the People's National Party (PNP), succeeding in establishing the Provisional National Defence Council (PNDC), a military junta functioning as a one-party government. (The party, later named the National Democratic Congress (NDC), governed Ghana until Rawlings lost the presidential election of 2000.) The 1970s had been an era of economic crisis across the continent. National debt skyrocketed, consuming the capital needed for development. A series of natural disasters had heavily impacted cocoa production and the Ghanaian export industry was nearly bankrupt by the early 1980s. Because these industries were essential to maintaining the Ghanaian government via tax revenue, and thereby essential to political stability, Rawlings adopted a policy of trade liberalization.[28]

Like Nkrumah, Rawlings saw culture as central to the nation-building project and essential to national economic progress. Also like Nkrumah, he carefully selected talented, politically like-minded artist-intellectuals who supported his vision, in this case a vision shaped by a populist ideology and a critique of neocolonial Westernization processes. One of Rawlings' more significant decisions was to appoint, in 1983, the Marxist playwright Mohammed ben Abdallah as Minister of Culture and Education. Abdallah deeply admired Nkrumah and was versed in his cultural agenda: he restructured the state's cultural policy and bureaucracy while drawing inspiration from Nkrumah's vision. In an echo of the Arts Council, he established the National Commission on Culture, which was composed of experts on many different aspects of culture. The Commission supervised all statal cultural institutions, including the Museums and Monuments Board, Folklore Board, National Theatre, and National Archives. It oversaw the ten restructured regional Centres for National Culture (CNCs), of which the Accra Arts Centre was the largest, as well as all parastatal cultural institutions, including associations for writers, dancers, musicians, artists, and so on. The Commission also supervised the new W.E.B. Du Bois Centre for Pan-African Culture, where Du Bois was entombed, and Kwame Nkrumah Memorial Park, where Nkrumah was entombed.[29]

Part of the PNDC populist ideology involved a thrust toward "democratization" (meaning a new type of inclusivity) and bureaucratization, which would come to incorporate "the whole of culture," as Nketia put it. It opened a space for art forms, which had been excluded by the postcolonial intellectual elite, to be incorporated into national culture. Whereas most Nkrumahists had concentrated on high art traditions, the PNDC gathered diverse craft traditions together to represent the nation of Ghana. Under this totalizing view of culture, Abdallah brought different craft-worlds that had been dispersed throughout Accra into the Arts Centre. He thereby rendered equivalent the distinct histories and products of Ewe woodcarvers, Hausa handicraft dealers, Akan kente cloth sellers and rattan weavers, and Dagomba basket-makers.

This reconfiguring of national culture transformed the Arts Centre, which had lost relevance for most Ghanaians even before Nkrumah was deposed. Scholars and artists have attributed the failure of Nkrumah's Euro-influenced forms of performance, music, and art to grab hold of the Ghanaian imaginary to their favoring passive consumption over active participation (Collins 2001). Amarteifio, for example, commented that Nkrumah "thought he could beat the Europeans at their own game. It was a mistake. We are not theatre-going people, so the Centre failed … It was only when theatre failed that the [craft] market and everything else came in."

The woodcarvers, who had been dispersed in workshops throughout Adabraka, were the first to come to the Arts Centre in 1979. They were soon joined by an even larger group of cane and rattan furniture-makers. Next to arrive came the kente dealers, who had previously been displaying their wares in front of the Ghana Library. Changes in the National Board for Small-Scale Industry, which had housed FraFra and Dagomba basket-makers, resulted in their shifting to the Arts Centre as well.

Not everyone was happy to be moved into the Arts Centre, for example many Hausa dealers resisted, because they depended for their livelihood on capturing the interest of passersby on High Street. As Lawan explained, state officials demanded that they move their kiosks from the street so the government could develop the Arts Centre:

> First, we were on High Street. Then when Rawlings came, he sacked us from there. [He told us to] bring the market [from] there to develop here, because here is a cultural center, but [there is] no anything there [on High Street]. So he said, we are cultural people. And they [government officials] do not like kiosk-kiosk on the street. They want all the people to be here.

To make room for this new, inclusive cultural vision, Abdallah expanded the Arts Centre east onto land that had been ignored by Nkrumah and subsequent regimes. Craft people at first resisted settling on this mix of forest, farmland, and Ga holy land, which many of them viewed as dangerous and outside the rule of law. While waiting to pick up an order from Lawan, a young assistant from Zongo Lane once told me, "Formerly, this was a huge forest. Robbers used to come over here to share their things. People used to shit. A whole lot of things [happened] over here." Ametewe described a wider range of activities:

> Before Abdallah, all the areas you see around, they are complete forest. Bush, big trees, people even farm along the places we pass through now…Where Ben Abdallah put the kente people, their structure, they planted plantain, cocoa, yam, everything. Even the market, the artifact market, it was farm.… It was a place where Accra [Ga] people practiced their festivals.…Here is the place where the chief fetish priest [would] do all the necessary customary rites before the festival starts.

Eventually, the craftsmen's resistance was overcome and the Textile and Handicraft Markets were built on the farmland. Carvers' Lane extended over the holy land, which was cleared by the craftsmen to make room for their shops. But the struggle did not end there, since these two parcels of land comprised valuable beachfront property. Instead, they became subject to a prolonged court case between the Arts Centre and the Ministry of Tourism. It was not until the early 2000s, under then-Director Alex Sefah Twerefour, that the land title was finally granted to the Arts Centre. And yet, even then, ownership of the land under Carvers' Lane remained unsettled. Threatening a future that never came to be, a faded sign advertising a never-completed hotel project stood over Carvers' Lane throughout the 1990s and 2000s. The sign symbolized for me the precarious impermanence that characterized the Lane and the neoliberal pressures which shaped it—the looming threat of removal or imminent destruction, and the sense of being unprotected and unappreciated by those in power.

As Ametewe commented in tones of pride and hurt:

> *We* are the people who make the name Arts Centre. So when we finally settle down here, it's there that the name become popular, because it's there that people come from outside the country, they come and buy from *us*. And then they go back and [are asked], "Where do you get this thing from?" "From Arts Centre, from Arts Centre." It becomes international. But not because of the administrative block [run by the educated elites]. Because of people like me, like ordinary people you see around doing their petty things.

Layer 6: 1990s Economic and Cultural Liberalization

The PNDC's aspirations were blocked by neoliberal changes to the Ghanaian economy as Ghana's deal with the IMF shrank the state's developmentalist promise. To receive international funding from the IMF and World Bank, the PNDC in 1984 agreed to a structural adjustment plan for cuts in social services and accelerated divestiture of government-owned enterprises. One of these conditions required the government to abandon direct control over and to limit its funding of its cultural enterprises. "Culture," which had become so heavily bureaucratized and central-ized during the early years of the PNDC regime, now had to be partially privatized and decentralized. As state funds flowed less freely to the cultural centers, the Arts Centre had to turn to what then-Director Twerefour termed "private partnerships" for funds. The Centre's cultural representatives—the performers, producers, and traders who had "made the name" of the Arts Centre famous—were turned into tenants on state land. One consequence of this change was that Centre adminis-trators increased the number of allotments, to raise more revenue. This led to a proliferation of shops in all sectors, heightening competition and stress. By the end of the decade, the original eight workshops on Carvers' Lane had expanded to 50, the original 18 Hausa kiosks that came off High Street had multiplied into over 100 shops in the Handicraft Market, and the Textile Market was similarly full.

Education and formal sector jobs and salaries were also cut under the stipu-lations of the international funding agencies. The socio-economic consequences would be broadly felt. The economic crisis cut into the heart of social life and caused a crisis in gender relations by preventing youth from obtaining the resources they needed to marry, start families, and thus achieve full adulthood. Young men routinely expressed that the state's promise of development and the implicit social contract between elders and youth had been broken by the 1990s.

Within the Arts Centre itself, the macroeconomic forces described above di-rectly impacted the relationships between older and younger men. As education became more expensive and formal sector work less viable, ever larger numbers of young men sought craftwork. Despite the proliferation of craftshops, the youth who came to the Arts Centre seeking work in the craft industries found few op-portunities to do so. Shop-controlling craftsmen were already deeply stressed, struggling to make a living, and unable to support any more dependents, includ-ing apprentices. Hence, without an option for employment in the craftshops, and with a proliferation of new shops opening up, all looking for customers, young men turned to hustling. This in turn drove tourists away from the Arts Centre, toward other, less challenging sites, where even "educated people," as Ametewe called them, who had formerly belittled the touristic arts were taking them up as a means of employment in the new economy. Simultaneously growing more crowded and emptier, the Arts Centre, in this way, would offer a material instance of the paradoxes of the Rawlings era.

Conclusion

The Rawlings era thus set the stage for the emergence of Rastaworld. The democra-tization of Ghana in the 1990s, attended by the liberalization and commercialization of formerly state-controlled media, ushered in a newly open public sphere. Once the state lost the ability to control culture, religion, and tradition, they would be taken up in a "marketplace of culture" (Marcus and Myers 1995) that was increasingly globalized by media flows (Appadurai 1991, Ferguson 2006, Meyer 1998, 2004, Piot 2010). With the Ghanaian state agenda focused on the convergence of culture and tourism in a privatized economy, images propagated through state media and state spectacles such as PANAFEST cohabited with a plethora of competing images, actions, words, and objects articulating new ideologies and possibilities (Keane 2003). In this climate, the "fluidity of musical consciousness as social identity" (Feld 1988:102) would remake Accra by simultaneously offering Ghanaian youth access to reggae, hip hop, and globalized forms of West African drumming.

Rastahood was born out of the dual necessity and possibility of the neoliberal moment, constituting in and through it a new form of tourist craft—informal, export-based, male- and youth-dominated. But it drew upon a broad array of resources produced at different moments over 500 years of history. The place where this transformation began was the Arts Centre. By the close of the millennium, the Arts Centre had become a holding place for various tactics for economic survival through cultural production. Walking through a multiplicity of moments—embodied in people, materialized in objects, emplaced in the land and built infrastructure—Rastas took up the Arts Centre method of selective use to create a cultural figure for the present moment, whether it be the Ga openness to exchange, an artistic dialogue with a colonial image of Africa, the Nkrumahist discourse of Sankofa, or the focus on tourism and diaspora of the Rawlings era. Rastas decontextualized and recontextualized, nationalized and then internationalized, ethnic traditions. Theirs would become an informal, bottom-up cultural tourism, which operated partly within the state's cultural and tourism apparatus and partly within the private sector. It would become not only a tourist art, which repackaged cultural heritage as commodifiable national culture in the global arena, but also a meaningful way of life. Rastas were a product of their time, but they were also the most recent example of an old pattern. Here, it had always been this: a combinatory logic of the new, brought about through contact, with the old, selectively repurposed and revised, to solve and transform the present.

Ghanaian youth understood, however, that it was also easy to become stuck there, barely earning a subsistence-level income, even as they sought mobility. One day in 2004, as we sat on a log bench near Akwaaba Restaurant, Noah said something that struck me as poetic, astute, sad, and true: "Everyone at the Arts Centre is there because they want to leave." A decade later, we were talking on the phone (he from his home in the United Kingdom, I from mine in the United States) and I quoted that line back to him. I don't know if I misheard him the first time, or if he misheard me the second, but in 2014, he heard me say, "Everyone at the Arts Centre is there because they want to live," and he agreed with his younger self. That both statements were right becomes clear in the next chapter focusing on the craft workshops of the Centre and the changing experiences there that necessitated Rastahood.

Notes

1 I draw inspiration from Steven Feld's concept of "lift-up-over sounding" (Feld 1988).
2 This co-extension of pasts differs from Jameson's famous theorization of postmodernism's flattening and commodified availability of pasts (1991).

3 Keith Hart first coined the term "informal economy" in 1973 to describe Ghana's nonwage labor sector (in contrast to its formal labor sector, defined by government-documented and -registered wage employment. The term is now frequently used to understand African economies. It includes workers, including porters, hawkers, market sellers, roadside sellers, drivers, artisans, and many tradespeople.

4 Ato Quayson's *Oxford Street, Accra* (2014) is a place-based history of Accra that resonates with what I attempt to do in this chapter. "Space provides the overall organizing principle for this book. I shall retell the urban social history of Accra from the vantage point of the singular Oxford Street, part of the city's most vibrant and globalized commercial district" (2014:4). For Quayson, Oxford Street enables an analysis of the "different spatial ecologies" generated under different regimes that explains the variegated and contradictory metropolis we find today (2014:4).

5 Parker (2000:xxiv–xxvi) points to such mixedness as the reason why Accra and the Ga present "a glaring lacuna in one of the most developed national historiographies of tropical Africa." The first modern monographs on Ga were written by Field (1937, 1940). Doctoral theses on Ga history have been written by Quaye (1972) and Quarcoopome (1993), while Odotei (1972, 1989, 1991) is considered the preeminent historiographer of the Ga.

6 For the history of the formation of Accra, see Abloh (1967), Acquah (1972), and Brand (1972).

7 For differing accounts of point of origin and merging with other groups, see Field (1940), Dakubu (1997), and Odotei (1977).

8 This can be seen in the way each Ga state comprised a number of *akustei* (quarters), each of which comprised several *wei*, the ancestral homes that are the basic units of Ga traditional society. The households (*shiaii*) are patrilineal bodies ostensibly tied together by a familial connection to a common ancestor, though this connection is more of an implicit claim of membership than an explicit line of descent (Quarcoopome 1998:134–135).

9 See Odotei (1991) and Robertson (1984) for other definitions of Ga Mashie. See Quayson (2013:Chapter 1) for a discussion of a "Ga ethno-cultural hybridity" (2014:44) in the 19th century, specifically of Tabon and Yoruba commercial networks within Ga Mashie.

10 A series of contiguous settlements developed into seven Ga towns arrayed along the coast from west to east: Nleshi, Kinka, Osu, La, Teshie, Nungua, and Tema (Robertson 1984:27). When the Portuguese arrived in 1482, the center of the Ga state was Ayawaso, a large town 11 miles inland. Little Accra (today's Central Accra) was still only a small fishing village (Odotei 1991:61). At the time, most of the coastal peoples were engaged in fishing, salt production, and trade.

11 Especially with its inland vassal state of Akwamu. In 1680, after a protracted war, the Akwamu destroyed Ayawaso and Little Accra (Robertson 1984:27). The Ga Mashie political structure fragmented, with some members of the confederation fleeing to Togo, while others stayed to rebuild Little Accra, which became the new capital of the Ga state. The Akwamu ruled the Ga until 1730, when a Ga-Adangbe-Akwapim-Akyem alliance ended the tributary relationship with Akwamu. Akyem claimed suzerainty over the Ga until 1742, when they were defeated by the Asante, who were then defeated by a British-coastal state coalition in 1826 (Odotei 1991:62).

12 The Anglo-Danish Treaty took place in 1850 and the Anglo-Dutch Treaty in 1874. The Northern Territories were incorporated as a protectorate in 1898, Ashanti was annexed in 1901, and part of (formerly German-controlled) Togoland was ceded to Britain in 1922, thus constituting the geography of the current nation-state of Ghana. For examinations of British presence on the Gold Coast and its effect on Accra, see Kilson (1974), Pellow (2001), Quarcoopome (1993), Robertson (1984). Statistics on the immense growth of this period are presented in Quarcoopome (1993:27) and Acquah (1972).

13 According to Quarcoopome (1993:22), Accra was selected over Cape Coast, Elmina, and Ada for its positional and climatic advantages: its dry plains and absence of mangrove swamps produced much lower malaria levels, and the absence of the tsetse fly made keeping livestock possible, two major British concerns. During the early colonial era, the whitewashed mansions of foreigners rose among the brown earthen huts of the locals. A lack of public latrines forced people to use beaches and surrounding bushes (Quarcoopome 1993:21).

14 The Towns Council Ordinance of 1894 gave the British Governor the rights to acquire Ga land, tax residents, and establish the Accra Municipal Council to oversee sanitation and maintenance (Brand 1972:64, Hess 2000:39, Sackeyfio 2012:298).

15 Meanwhile, a series of natural disasters between 1890 and 1919 weakened Central Accra and resulted in ordinances that enabled the British to further remake the city, resettling people from the increasingly overcrowded Ga quarters into the new suburbs of Riponsville and Adabraka (Hess 2000:40). In response to a bubonic plague epidemic in 1908, the government issued the Infestation Disease Ordinance and, in conjunction, created a Demolition Committee, which tore down many of the homes in Central Accra and reorganized the layout of the town. This has mixed results, since local residents, who mostly subsisted on fishing, opposed the demolition. Many then refused the government's offer to move to the new suburbs, claiming these were too far from the sea.

16 Introduced in the 1850s as a cash crop (Herppich 2016:189), cacao rose drastically in importance throughout the British tenure in the Gold Coast. This can be seen in export volumes reported by the colony, where cocoa exports rose from roughly 500 tons in 1900, to 1,000 tons in 1902, to 236,000 tons in 1937 (Annual Report of the Colonies, Gold Coast 1902:180, 1936–37:128).

17 Prior to the interwar solidification of rule, education had been the domain of missionaries and, through to the end of the 19th century, was of the classical sort: African schoolboys learned much what Charity School English schoolboys did (King 1971:44). Already by the 1890s, what was "variously described as the problem of the educated African, the over-supply of clerks, the mission boy, or the Black Englishman" (1971:45) was already noted. Complaints were coming from missions throughout the continent. The famous travel writer Mary Kingsley "affirmed in the mid-1890s that there was 'no immediate use for clerks in Africa'; all Africa would need for the next two hundred years at least was a supply of 'workers, planters, plantation hands, miners and seamen'" (1971 [1897]:46). A new kind of education was needed which would form the African for his place in the imperial world order. In a sarcastic yet apt portrayal of the governmental position on African education, the American educator Wallbank states: "The most important object of native education … is the smooth and effective functioning of the governmental system. Education should be directed so as to prevent the natives from dallying with any 'disturbing' ideas concerning self-determination. … the students should be taught to be loyal subjects, content to 'know their place.' The government should provide only limited opportunities for higher education such as are necessary to produce a small intellectual elite from which can be recruited the government clerks and puppet chiefs who act as useful links between the white rulers and their native subjects" (Wallbank 1935:232).

18 This curriculum was imported from the United States in the form of a 1925 British colonial White Paper entitled "Education Policy in British West Africa." This paper "was almost an exact replica of the Phelps-Stokes report of 1922," with which Guggisberg was already acquainted, and which had quickly become the authoritative document on colonial education (Ofori-Attah 2006, Steiner-Khamsi and Quist 2000:273). The Phelps-Stokes Foundation, a New York-based philanthropic organization established in 1910 with the mission to research, fund, and otherwise support Negro education in America and Africa, promoted an industrial and trade-oriented education as an "antidote against

the corrosion" of the Pan-Africanism then being promoted by African American activist-educators such as Marcus Garvey and W.E.B. Du Bois (King 1970:18, Omolewa 2006). In 1899, Booker T. Washington came to Britain to present Alabama's Tuskegee School as an example of industrial education, thus providing a welcome new alternative to missionaries and colonial educators dealing with the African education problem. This visit along with subsequent publications and correspondence led Sir Michael Sadler, well-known educationalist and Britain's Board of Education's director of special inquiries, to visit Hampton Institute. In 1902, Sadler became the first Briton to urge the transfer of industrial education to Africa (Steiner-Khamsi and Quist 2000:296). His focus on the arts was so strong he edited a volume titled *Arts of West Africa* containing essays by Achimota's art tutors (Sadler 1935, see Whitehead 2003).

19 See Antubam (1963), Coe (2005, 2014), Harrod (1989), Kwami (2013), and Shipley (2015) for analyses of the specific role of performance, music, arts and crafts in Achimota education, and Gilvin (2014) for colonial art education more broadly. See Botwe-Asamoah (2005), Coe (2005), Schauert (2015), Shipley (2015), Seid'ou (2014a, 2014b) for Achimota's centrality in the formation of Ghanaian ideas of culture.

20 Many of Ghana's national artists passed through the art teaching program, for, as artist Ablade Glover told me in 2004, "back then, you could not make money from art, you had to teach." Financed by the British Colonial Development Fund, Meyerowitz then established the West African Institute of Art, Industries, and Social Science; he there expanded the arts curricula to include many kinds of crafts (Antubam 1963:200). Upon his death, his wife Eva, a sculptor and ethnographer, ran the art department for an additional year. The industrial and science programs were transferred to the University of Science and Technology in Kumasi in 1952, while the performing arts were transferred to the University of Ghana at Legon.

21 The Hausa Constabulary was established in the 1860s by Captain John Glover to protect the Lagos colony. Part of this constabulary, consisting of several hundred Yoruba and Hausa, many of them runaway slaves, was sent to the Gold Coast to fight in the Anglo-Ashanti War (1873–1874), to be renamed the Gold Coast Constabulary in 1876. Four hundred of its members were recruited as police officers following the Police Ordinance of 1874. Again, although named Hausa, this group of soldiers and police officers were in fact from disparate ethnic groups from the Northern territories, but they all spoke Hausa and were thus essentialized by the colonial rubric (Quayson 2014:192–194).

22 According to the native law of the Gold Coast, "a person of a different tribal origin is regarded as a stranger" (Pogucki 1954:31). Strangers could not aspire to own "stool" land, but Ga policy toward Northerners was generally welcoming. Early Northerners entered into patron-client relationships with the Ga (Pellow 2002:47). They had fewer rights in the European colony, however, where the British viewed them as foreigners mainly to be used as counterweights to Ga power (Pellow 1985:432). As the city grew and land gained in value in the 20th century, "arguments about primogeniture became central for adjudicating land ownership, [and determining] who had natural claims to the land and who could claim ownership only as a stranger" (Quayson 2014:39).

23 The independent state of Ghana retained Elizabeth II as its queen, while Kwame Nkrumah was its prime minister. The 1960 constitution replaced the Queen with a President (Kwame Nkrumah) and Ghana became a Republic within the British Commonwealth of Nations.

24 In practice, attempting to foster Pan-African socialism for the long term while safely accessing the capital of former colonial rulers in the short term was difficult. Economically, Nkrumah therefore advanced a mixed ideological policy, featuring both public and private sector industries with a guiding hand from the government. Although private interests were allowed to develop in the socialist state, what is important to note is that only foreign capital was permitted to be held privately in any significant amount.

Nkrumah valued foreign investment and maintained cooperation with former colonial powers but was wary of allowing private businesses to become entrenched in the Ghanaian economy (Apter 1972).

25 The extensive literature on the role of the arts in African nationalist projects includes Askew (2002), Castaldi (2006), Edmondson (2007), Harney (2004), Turino (2000), Winegar (2006). For Ghana specifically, see Allman (2013), Coe (2005), Fuller (2014), Hess (2000, 2003), Lentz (2001), Schauert (2015), Shipley (2015), Wilburn (2012).

26 See Fanon (1965), Mudimbe (1988, 1992), and wa Thiong'o (1986) on elite constructions of African culture.

27 From 1966 to 1969, the National Liberation Council (NLC) ruled Ghana, during which time it drafted a new constitution and formed the Second Republic of Ghana. Only in 1969, were nationwide elections held, the first since 1956. They put Kofi A. Busia and the Progress Party (PP) in charge for a little over two years. Then, on January 13, 1972, Lieutenant Colonel Ignatius Kutu Acheampong ended the Second Republic through a bloodless coup, establishing the National Redemption Council (NRC). Unlike the NLC, the NRC did not seek a transition to democratic rule. Instead, it was reorganized in 1975 into the Supreme Military Council (SMC). This military government ruled Ghana until June 1979, when a constitutional government was supposed to be reinstated. However, on June 4, 1979, Flight Lieutenant J. J. Rawlings led the Armed Forces Revolutionary Council (AFRC) in a bloody coup, replacing the SMC. The AFRC then allowed the scheduled elections to take place and handed power over to the People's National Party (PNP), with Dr. Hilla Limann becoming president of the Third Republic of Ghana.

28 Rawlings began with a complete restructuring of the tariff system, drastic reduction in civil service employees, and implementation of multiple Structural Adjustment Programs mandated by the International Monetary Fund (IMF). Rawlings' economic plan included two distinct phases intended to, first, increase the profitability of existing exports and internal industries by cracking down on the black market and, second, trim government regulations and increase farmer profitability while eliminating informal trade in cash crops for good. The second phase included the development of "Export-Promotion Villages" and alternative export commodities. Under this program, the PNDC provided grants and expertise to Ghanaians who could produce art objects, woodcrafts, and non-cash crops; these products were expected to offer more stable sources of revenue for the Ghanaian state (Grant 1999).

29 Rawlings had built the mausoleum to continue the rehabilitation of Nkrumah's image, begun under General Ignatius Kutu Acheampong in 1975. Nkrumah had died in exile, but Acheampong had Nkrumah's body returned to Ghana and buried in his hometown. Rawlings had Nkrumah's body unearthed and brought to Accra, where it was entombed him in a mausoleum in the memorial park.

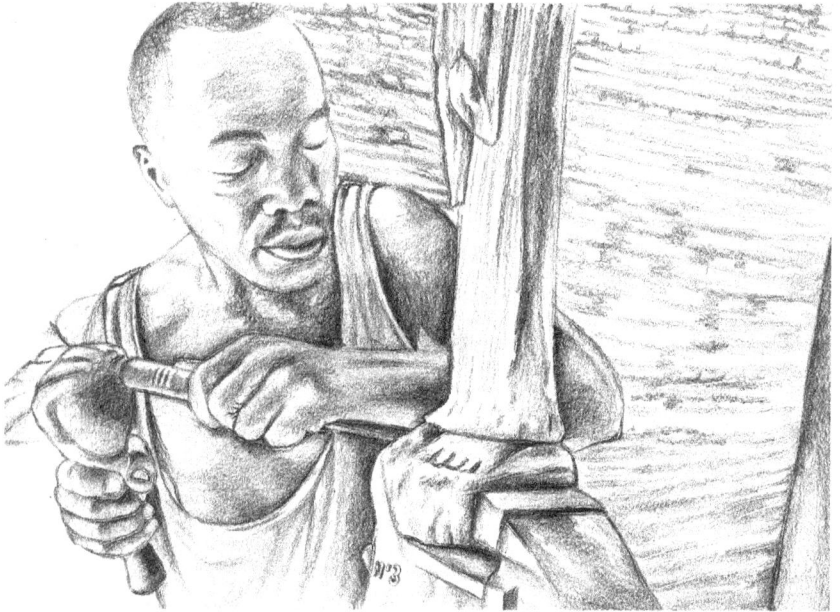

2 Men at Work

Craftwork, Masculinity, and Precarity

To apprehend Rastahood, one needs to understand the changes to craftwork that took place under neoliberalism. While the Arts Centre was a site of possibility that generated Rastahood, it was also representative of the obstacles that led youth to create the form. For several decades, the Arts Centre craftworld provided working-class men with viable livelihoods, enabling them to fulfill cultural scripts of adult masculinity in Ghana. But by the late 1990s, neoliberal economic changes made fulfilling these scripts impossible. This affected both men, who found themselves losing their foothold in the new economy, and youth, who could not depend on men to help them, creating a crisis in intergenerational relations. Together, the script, its impossibility, and the crisis shaped how and why youth became Rastas. In this chapter, I bring together craftwork, masculinity, neoliberalism, and time to explain why Rastahood emerged in its particular form and at its particular time.

Like Rastahood, urban craftwork for the tourist trade was not the ultimate desired destination for Ghanaian youth. It was a means to an end: as Noah said, everyone was there because they wanted to leave. Unable to access higher education and the resulting economic and professional success, men entered craftwork in the hope that their children could achieve what they had not. But for a time, craftwork, specifically the craft workshop, offered a (delimited but nonetheless real) solution to various challenges in men's lives; it was a hinge between the social and capitalist time, a way to advance the next generation, and a space of respect and masculine authority. Key to this was the temporality of the workshop and of craft itself, in which slowness provided time for adulthood and artistry. This chapter traces the work model that Rasta youth inherited, and then the politico-economic changes that made achieving it impossible. As the passage from present to future, and from boyhood to manhood, became increasingly contingent, youth in Accra were forced to expand the spaces of exchange and opportunity. Adulthood could no longer be attained by remaining in one place but rather must be chased by making connections over ever greater distances. Rastahood was thus a "spatial fix" (Mains 2007) to the problem of neoliberal time.

Craftwork was an ambivalent category of labor and identity in Ghana throughout the 20th century. As I discuss in the previous chapter, for Europeans, handicrafts such as woodcarving, basket weaving, and metalwork materialized "Africa" as a place of unchanging, collective, rural traditions imbued with spirituality. African crafts were a soothing antidote to the modernity of home, with its attendant

DOI: 10.4324/9780429244568-3

urbanization, individualization, and uncontrollable pace of change. They were also a soothing antidote to modernity in the colonies and the implications of a European loss of control there. Craftwork also became a site of class conflict and postcolonial angst amongst elite Ghanaians (Mbembe 2001). Its colonial associations with rurality, manual labor, ahistoricity, and reified tradition made craftwork antithetical to cosmopolitan modernity.

For non-elite men, making crafts for the tourist market produced a different set of valences. While seldom their first career choice, from the 1960s through the 1990s, urban craftwork (and the shops where it was learned) constituted a zone of dignity for Ghanaian men. Unlike other forms of manual labor, such as construction, portage, or hawking, craft was creative work, and the workshop was a space of authority and ownership that produced relations of respect. For youth who lacked higher education or elite networks, and thus access to the formal sector, careers in urban craft promised an alternative track to financial independence and its concomitant social reproduction of gender- and age-based identities and relations. That is, financial independence translated into the support of dependents, both familial (wives and children) and professional (apprentices), which defined adult masculinity. Craftwork provided dignity when dignity was hard to come by.

While they had to work steadily to produce and sell enough carvings to pay their bills and support their dependents, craftsmen found a delicate perch in urban craftwork, which enabled them to suture capitalist time to social time. With money flowing in from the tourist trade, men could transmute their labor into social reproduction and negotiate the conflicting demands placed upon them by dependents and expenses of urban life (Achebe 1960, Pedersen and Nielsen 2013). The workshops at the Arts Centre served as transtemporal hinges (Pedersen and Nielsen 2013), linking capitalist time—an abstract, empty time to be filled with production value—with social reproduction in a particular locale (Bear 2016). In this way, the life of carvers at the Arts Centre Modern reflected the heterochrony of time-maps that shape people's experiences and desires and constrain their agency in heterogeneous ways (Bear 2014, Gell 2001, Munn 1992). The social time-map produced a linear progression through the life stages and cycles of commitment to kin, entailing payments of school fees, medical bills, household expenses, and funerary gifts. The capitalist time-map, closely linked with urban life, produced the abstract, requisite rhythms of production, circulation, and return, accompanied by impersonal demands for payments of rents, fees, and bills. The space of the workshop mediated "the clash of diverse and antagonistic temporalities" that defines capitalism (Negri 2003:68).

Throughout the latter 20th century in Accra, the craftshop was a space where elder men transmitted patriarchal, gerontocratic ideals and social relationships to their apprentices, which ensured the reproduction of the social system from generation to generation. Age hierarchy, exclusion of women, and ethnic control over specific crafts constituted forms of power for senior craftsmen. Moreover, the intersubjectivity that characterizes "work in the workshop" (Förster 2013), while transmitting ideas about art, labor, and value, also defined gender through iterative practice (Butler 1988, Sapir 1949). As long as it remained embedded in a stable economy, the workshop system thereby effectively provided non-elite men with a living, an identity, a sense of place, and a means toward achieving social adulthood.

Senior craftsmen embodied and transmitted the social schema working-class youth hoped to adopt so they in turn could become "men" (Johnson-Hanks 2006). That is, through learning a craft, these youth could expect over time to become shop masters and heads of urban households able to educate their own sons and move them further on the path of upward mobility.

By the 1990s, the craft system within the Arts Centre had become solidified, with the ethnic enclaves that dominated particular crafts functioning as landing places for new arrivals to Accra. Most of the newly arrived youth came from Ghana's rural hinterlands, where their families were struggling to support themselves. Some were political or economic migrants from neighboring Togo, Burkina Faso, Mali, and Ivory Coast. Through various kinship, religious, or ethnic networks, many would come to seek entry into the Arts Centre's workshop system. These youth usually hoped to enter one of three well-established tourist trades: handicraft dealing, a predominantly Hausa or Muslim trade; basketweaving, which was done in familial workshops controlled by FraFra men from the Upper West Region of northern Ghana; or woodcarving, controlled by Ewe men from the northern Volta Region. Why, then, at the end of the millennium, did they invent the new craftworld of Rastahood?

This chapter addresses this question by investigating the models of art, labor, and personhood into which youth were being socialized at the very moment when such models were ceasing to be viable. It begins by exploring the lives of some of the first craftsmen to move to the Producer Section of the Arts Centre known as Carvers' Lane. While these men happen to be woodcarvers, their production of a respectable, aspirational working-class masculinity through craft labor and then their struggles to do so in neoliberal time are representative of the predicaments of other working-class men at this time. As Ghana shifted toward neoliberal capitalism, the sense of time in which craftsmen labored became altered. The long-term apprenticeships and slow work of woodcarving could no longer be sustained. Youth arriving in Accra at the turn of the millennium were forced to seek or invent new forms of labor if they hoped to survive the constantly changing pace of neoliberal time.

Crafting Masculinity in Carvers' Lane

Youth who joined a woodcarving workshop between the 1970s and early 1990s anticipated having careers that would unfold like those of the senior craftsmen. Following an established cultural script, they are expected to enter the workshop as a space of professional, cultural, and social knowledge transmission. They would begin as apprentices and, after three to six years of hard work, loyalty, and learning, become masters of their own shops (Peil 1970). Learning to labor in this way was inextricably tied to attainment of adult status and respectable personhood. An apprenticeship was a forward and upward moving trajectory from childhood, through the transitional category of youth, and into adult masculinity (Cole 2010, Comaroff and Comaroff 2001). This professional transformation moved youth from the social category of a dependent "boy" to that of an independent "man," capable of supporting dependents of his own.[1] In the intersubjective space of the craftshop, apprentices (dependents) and masters (their supporters) co-produced a proper adult masculinity.

Godwin Ametewe: "Carving Is a Complete Institution"

When I first sought permission to do research in Carvers' Lane, I was directed to the workshop of Ewe woodcarver Godwin Ametewe, the Secretary of the Producers Association and historian of the local handicraft trade. His open-faced shop was cool and quiet, a refuge from the sun that blazed onto the sandy lane. Just inside the entrance, a small table displayed elephants ranging from 7 to 30 inches high and carved out of mahogany, rosewood, or ebony. Such carvings of "African" animals, most of which he and the other carvers had never actually seen in the flesh, were intended for foreign tourists. Against one wall stood a standard carpenter's bench, which he always dusted, removing invisible dirt, before inviting visitors to sit down. His worktable was at the back of the shop behind an enormous mahogany hippopotamus (another iconic animal form) that also served as a bench. Over time, the hippo had acquired a deep burnish from all that sitting, a glowing patina of time. Like everyone else in Carvers' Lane, Ametewe stored valuables such as tools, wood polish, and sandpaper under tarps below the tables, hung from the walls, and tucked into the rafters. His shop was the *de facto* general office of Carvers' Lane, where he also maintained careful records of the ever-shifting ownership, occupation, and division of the shops, ran general, special, and mediation meetings, and oversaw the entire Carvers' Lane community, including the other Ewe woodcarvers, FraFra basketweavers, Hausa and Mossi handicraft dealers and finishers, and diverse instrument makers and leatherworkers.

Ametewe was born in 1957, the year of Ghana's independence, in a village in the northern part of the Volta Region. He was the seventh of nine children, the youngest of three sons. The entire family farmed for a living. His father, like the other men in the community, also carved wood stools, spoons, drums, toys, and other functional objects to sell to supplement the family income. Ametewe had a brilliant, incisive mind. Given the opportunity, he would have become an agricultural scientist, but his family could not afford to send him to secondary school or university. And so, on a visit home, Ametewe's older brother, who was apprenticing in Accra with a village carver, convinced Godwin to leave the village and its lack of opportunity and return with him to the capital. He promised to introduce him to an American man who might sponsor his education. Ametewe recalled:

> I actually did not know that one day I would be a carver; carving was not my preferred profession. So when I was coming with my senior brother to Accra—then, he was working with an American man in the American Embassy, he's called Lee—my brother promised me that, when we arrive in Accra, he can introduce me to him so he can help me to go to secondary school. So I decided to come, but it was not possible for the American friend to grant the request of my brother. So I, well, I left the village. I cannot go back to the village, too, because I know what pertains in the village. So what else do I do? I decided to stay with my brother, to help him. So in the course of time I developed an interest in the work. So I decided to learn, and that's how I became a wood carver.

Having left his village for Accra, without finding the opportunity to acquire the formal education that he sought, a craft apprenticeship became Ametewe's best alternative.

Entering Accra in the 1970s, Ametewe joined a community of carvers who had emigrated to the city from a cluster of neighboring villages in the northern Volta Region. They were united by shared culture and language as well as kinship relations, cross-apprenticeships, and craft practices. The Ewe carver network constituted an alternative family and support system away from the village and provided migrants with a landing place in the capital. Every carver or apprentice had been given a mental map of the city, including the workshops where they might work, hold meetings, sleep, store their things, or leave their goods to be sold when they left for home (Lynch 1960).

As carvers reported, most youth who became apprentices were introduced to a master carver by a family member. An elder would officiate a ritualized negotiation involving payment of money, schnapps, and soft drinks to the carver, after which the carver would take on the boy as his apprentice. Communications and learning processes were governed by the acknowledged expertise of the senior men. Boys were expected to learn "through others, not from them" (Kasfir and Förster 2013:13) in a situated process that involved hanging around, observing everything, obeying all commands, running simple errands, and eventually practicing the more complex tasks of carving (Lave and Wenger 1991).

These master-apprentice relationships constituted a familial network beyond consanguineal kinship, reflected in the language of the workshops. An apprentice, regardless of age, was called a "boy" to emphasize his dependency on the "masters," senior men also known as the "fathers" of the carving trade. In turn, as Ametewe described, senior men were obliged to look after their apprentices:

> If you have been able to pass the true process, you become a family people. … Take for instance that John [another master carver] is not here, and the boy he trained, the person he trained [in the village], comes to meet me in Accra. And I ask him, "Who is your master?" And he says, "John is my master, therefore, I'm begging you to help me." Because I know John, I will accept him so that he could be able to make some few cedis [Ghana's currency]. … He [John's apprentice] also takes me as the "father" because the father [John] knows me. … So their problem is my problem or our problem [and] our problem is their problem. So we do not have difficulties about identifying our people, we do not have that difficulty at all.

Thus established as a relationship between a "father" and a "boy," apprenticeship served as a period of transition that gave youth the tools they needed to cross over into adulthood.

Master carvers taught youth about the materials of carving, including the different kinds of hardwoods, where to procure them, and how to dry and store them, as well as the tools and techniques of carving, and the various genres of the products. Ametewe explained, "There are structures we follow before [an apprentice] could be called a carver. And most of us who have followed that structure, we will always excel at everything, because carving—it is somehow a way of life. There are a lot of steps—from here you go to here—before they call you a carver. It is not just a mere thing." In teaching their apprentices the craft, the senior men also transmitted "a way of life": cultural knowledge, ethics and morality, and information about how to run a business. Again, Ametewe elaborated:

> Carving is a complete institution. When you've gone to learn carving, it's your livelihood. So they will train you—how you get involved with your business, *how* you get involved with your family when you are married, *how* you are able to break up with your family, how you will be able to send your children to school, how you will be able to cope with the extended family, business practice, leadership qualities—they will teach you *all* these things. They assess you and grant you as a carver.

Indeed, Ametewe understood that the workshop system was just as much a "complete institution" as any profession that requires youth to become formally accredited (Boampong 2015), drawing as it did on "a stable framework of communication and learning governed by the acknowledged expertise of senior members of the group" (Kasfir and Förster 2013:2).

> If you are university graduate, if you practice law, it doesn't matter where you graduated, the moment you present your certificate, everybody knows that really, this man is qualified to be a lawyer. So what we do is, if you have been able to mention the place where you practiced or you learned your profession as a carver, because we know all the points, the areas where you practiced and learned as a carver, then we know you hold the banner of the profession. So people will say, "This man is my master. I passed through his stewardship." So you definitely know that that man is trained as a carver, so you accept him. As [far as] the work ethic is concerned, we are from the same family.

Of the many craft trades available at the Arts Centre, woodcarving most epitomized the dignity that could be attained through craftwork because it was the hardest craft to master. Wood was considered the most exacting medium and acquiring the ability to carve it took much longer than achieving proficiency in other crafts. Carving was also considered to require the utmost creative prowess. It was an act of the imagination, the translation of a mental image into form via the mediation of material, discipline, and will. Moreover, carvers such as Ametewe often explained that they had to see the object inside the wood, in a kind of God-given vision. "You see," Ametewe said, "Ewes are people of spiritual background, so we were doing this carving to represent our gods. … So carving, per se, it's with us, because … if the god revealed itself to you in a physical form, … you must go and look for that object, you must go and look for wood to carve in the same object." As an attempt to replicate, translate, or materialize sacred images, carving connected man to divinity, with each man's vision, knowledge, and life force being objectified in wooden form.

Of all the crafts, carving thus materialized a model of art as agency (Gell 1998). Carvers lived with carvings from their beginning, as mental images and hunks of wood, to their end, as polished objectifications of their vision and labor. Such productive activity was thereby a means for men to express their power, to leave their mark upon the world. Thus, within a capitalist market oriented toward producing tourist objects and controlled by outside mediators, carving, which required skill, time, and an end-to-end process, was less alienated than other kinds of labor available to working-class men in Accra in the latter 20th century (Kasfir and Förster 2013). In the workshop, the relative powerlessness carvers had in the art market was situationally offset by their senior role in traditional hierarchies. Moreover, the double act of imagination that underlay carving—imagining something invisible and then imaging how others would perceive it (Kasfir and Förster 2013:5)—afforded a way to reframe the power dynamic between carvers and consumers. Carvers at the Arts Centre understood that they were being reified by foreigners. But they also found that they could rise above their own objectification by viewing carving as agency, commission, and translation (Kasfir 2007). Ametewe continued:

The things that we see around ourselves are the things that we carve. ... So when the Europeans came in, they developed an interest in our culture, our music, our dance, our everything. The way we eat. The way we sleep. The way we put up our buildings. The way we wear our clothes. You see, they developed an interest in these things. So we also take advantage of that. Then we were putting our art in those types of objects so that when they buy it, they will keep on remembering what we were doing. So that's how commercial carving per se started.

Thus, the craft carved out a narrow niche from which men could feel like men, but neoliberal pressures, experienced economically, existentially, and temporally, shrunk the niche down even more. Manhood took time, and it took money, and these were incommensurable exigencies. In a voice filled with anguish and anger, Ametewe explained:

Do you know a lot of women left their husbands? Do you know? You do not know? You see, in our tradition, the husband is supposed to provide everything in the house, the recreation, food, entertainment, anything. So, and the woman's duty is to look after the house, after the children, do this and that, prepare the food. And you—the man in the house—you are supposed to provide all the money for all these things. And you are not able to do it, you cannot feed your children, you cannot feed your wife, you cannot pay school fees, you cannot pay your rent. That your landlord will have to throw you out of the house. Where do you think you will stay with your wife and your children and their schooling? So in the process of that, you have to send your wife and the children back to the village to stay with their families. And when they go there the schooling system is not good, so it means you are walking backward.

Having moved to the capital, site of modernity, opportunity, and progress, to earn money to advance himself and his family, a man now had to "walk backward" to the village (Ferguson 1999).

Mawuli Awoonor: "Carving Takes Time"

Like Ametewe, Mawuli Awoonor, widely considered the greatest artist in Carvers' Lane, approached carving with the utmost respect. As was typical of all the workshops on the Lane, a small table displayed his finished wares at the entrance to the shop and there was a bench where his visitors could sit whenever they came by. A 3-foot tall, solid mahogany elephant, which he carved, stood at the entrance for years gathering patina, until it finally sold in 2005, leaving a strange emptiness in its place.

I saw Awoonor manifest all manner of things from raw pieces of wood. For example, "elephant sets" (sets of five small carved elephants in incremental sizes) and elephant chairs (large elephants with flattened backs), from which Awoonor made most of his money, selling these to tourists, foreign sellers, or market traders. But he could carve anything, turning out a perfect shoehorn one day and a hippopotamus the next. And he was known, and deeply admired, for carving graceful, complex human figures, usually women with jars on their heads, babies on their backs, or baskets in their arms (a classic genre of Ghanaian art). Such figures were the hardest, took the longest, and were least profitable. Awoonor usually carved them uncommissioned, hoping each piece would eventually sell. Sometimes he carved for no payment at all.

My favorite memory of Awoonor at work is of two quiet, shy Ga children—a boy and a girl, roughly eight and ten years old—entering his workshop the week before the Ga Mashie Twins Yam Festival, an important pre-celebration rite of *Homowo*, the annual Ga festival (Fosu 1999, Kwakye-Opong 2014, Lentz 2001, Opoku 1970). The children, timid to the point of speechlessness, held out to him a badly carved wooden figure needed for the festival. They had been sent to the Arts Centre from Jamestown to have it fixed. Awoonor was ambivalent about taking on the job, but once he agreed to do it, he brought his considerable skill and respect to the task. With tiny flicks of his chisel, he rebalanced the proportion of the body, drew features out of the hacked surface of the head, rounded the limbs, and articulated the hands, feet, and hair. The children and I watched in silence as a specific, recognizable person emerged from the small, rude form.

In general, Awoonor's shop was much busier than Ametewe's, because he shared it with two other carvers, a part-time apprentice named Michael, and an auto mechanic who moonlighted as an itinerant carver's assistant. All five men hailed from neighboring villages in northern Volta. Between them, something was always being made or sold, and someone was always stopping by to visit, commission, or buy. The three carvers shared a massive worktable, but each had his own corner and worked standing in front of his own clamp. Another worktable filled the back of the shop, alongside the ubiquitous, wood-dusted, tarp-covered mounds of materials and goods. Here, the assistants energetically polished the carvings with ever-finer grades of sandpaper. Not only did each carver work separately, each also had his own style and genre specialization. Each's hand was imprinted on the surface of the wood before it was sanded down, and their artistic styles marked their piles of wood shavings as clearly as if they had named them. Each also had his own rhythm, and the different beats of their mallets hitting their chisels syncopated with the music from the radio and the sound of the assistants sanding wood.

Awoonor's own relationship to the wood was meditative, each unwasted, practiced stroke further exposing the object. In one hand, he held an ebony mallet he had carved in the shape of a fist holding a rock. In the other, he made a path through the wood with the edge of his chisel. One day, I watched his precise blows carve a woman carrying a basket out of an unworked hunk of ebony fastened in his clamp. He had assessed the wood before clamping it down, mapping the most economical number of moves it would take to carve the figure. The long, smooth, uninterrupted wood shavings, thick and solid at one end and curling thin at the other, gathered at his feet, knowledge objectified. It had taken years of apprenticeship, decades of carving, hours of thought, and a lifetime of studying, selecting, and drying wood to achieve such expertise.

"Carving takes time," Awoonor said. Time was a measure of his dedication, morality, and discipline, his respect for his art and for himself. The longer it took to produce an object, the greater its artistry, the more could be charged for it, and the more respect accrued to the master carver. Time thus produced artistic, economic, and social value, but it became increasingly scarce. In Awoonor's workshop, these changes could be seen in the greater success of Samuel, the youngest of the three carvers. Unlike Awoonor and Josiah, an even older carver who carved time-consuming, elaborate figurative works, Samuel, the youngest of the three, had best adapted to changes in the market. He worked closely with dealers, who commissioned large sets of ebony antelope-shaped bookends or stool-shaped napkin holders. He carved with incredible speed, churning these simple forms out with almost machine-made precision even though he measured everything by hand and eye. Samuel produced 20 or 30 objects for each carving by one of his elders. And yet, even his quick, whip-like moves, which left an almost invisible mark on the wood and piled up arrow-like shafts of shavings, could not keep up with the demands being placed on commercial carvers.

Even more telling, I once ran into Mishael, Awoonor's apprentice reselling another carver's horses to a dealer who had a shop in the Handicraft Market. He had recently returned to Accra from a trip to his village, and I had seen him sanding these carvings in Awoonor's shop a day or two earlier. I mistakenly assumed a village carver had made an agreement to display his wares in Awoonor's shop and paid Michael to sand and polish them, a common practice. Michael had purchased the horse carvings in the village, finished them himself, and was now arranging for them to be sold at the dealer's shop. After subtracting his expenses (including transport, purchase, and his own labor) and the 15% of the final sale that went to the dealer, Michael would earn less than $1 per object. But even this was an improvement over the slowness of a carving apprenticeship. "I want to go into business," he explained. "Buying and selling is faster than carving." With this small change, he was moving out of the workshop and into the market, shifting into the hustle of neoliberal time. It was an admission that the time-consuming nature of carving—the very element that bestowed its dignity upon men—was no longer a viable occupation for youth of the day.

From the Linear Future of Craftwork to the Circular Present of Neoliberalism

In Ghana, and indeed throughout the world, a shift in labor practices and the experience of time was propelled, beginning in the 1980s, by the rise of neoliberal capitalism. Defined by privatization, waning welfare state protections, and deregulation and casualization of the labor market (Martin and Ross 1999), neoliberal economic policies replaced long-term, protected employment with tenuous, insecure, short-term, contract work (Ong 2006, Sassen 1999, Sennett 1998). Ideally, the neoliberal worker was expected to earn a good living by being "flexible" and responsive to market demands (Freeman 2000, 2007, Harvey 1990, Sassen 1999), but in reality, irregular piecework, menial "workfare," and transient, profitless occupations have generated conditions of precarity for most millennial laborers (Comaroff and Comaroff 2001:5). In turn, since labor is inseparable from subjectivity, the precarity of temporary, irregular work increasingly ushered in existential states of anxiety, risk, and unbelonging (Millar 2014).

Precarity also fundamentally altered the subjective experience of time, rendering the progressive sense of time as forward movement and upward improvement, which characterized Euro-American modernity and development narratives, increasingly difficult to maintain. Throughout most of Africa, this meant that the promise of a better life following independence from colonial powers collapsed as these economic conditions took hold (Ferguson 1999, 2006). The near future was instead evacuated of utopian expectations and became radically unpredictable (Bear 2016, Mains 2007). The social changes that came with the global neoliberal economy made long-term planning, and the ability to see a causal link between intentional actions in the present and desired results in the future, increasingly elusive (Ferguson 1999, 2006, Johnson-Hanks 2006). People's sense of time thus shifted from a linear present-to-future movement to an ongoing present often experienced as either boredom or crisis (Masquelier 2013, Ralph 2008, Roitman 2013, Weiss 2009). The seemingly eternal present of neoliberal time came to be viewed as an emergency precisely because nothing was happening to suggest the possibility of change (Berlant 2011).

In Carvers' Lane, the transformations of work and time demanded a new approach to craftwork, that it be quick to learn and quick to do. Since woodcarving was neither, carvers were hit particularly hard. They either had to change their way of working or lose their livelihoods.

Afrane Dawson: "The Work Itself Is Lost"

Afrane Dawson was one of the few men who made a successful transition from carver to crafts dealer. A self-defined "Northerner," Dawson, was born in 1970 in Tamale, the capital of the Northern Region, to an Ashanti father and a Komba mother. When he was 12 years old, his father died and his paternal relatives took him and his siblings from their mother to live in the Ashanti family home in the Central Region. Although he had a home with one of his cousins, Dawson had to pay his own school fees and other expenses. He demonstrated an early talent for art, so an art teacher urged him to go to Accra to sit for examinations and complete secondary school with an art specialization. When his attempts to procure support for this schooling from elder kin failed, he landed at the Arts Centre and started working as a carver's assistant. For several years, he slept in his master's shop instead of struggling to pay rent elsewhere. Having been taught by the first generation of carvers to come to the Arts Centre, Dawson occupied an interstitial position between the senior men and incoming youth. His early life trajectory otherwise mirrored that of many other young men who ended up in the urban tourist craft trade.

Dawson loved watching the carvers work and quickly internalized the values of their craft. Thanks to natural talent and extreme self-discipline, he soon became a good carver. In local parlance, he was "fast"—he picked things up quickly and pivoted to new positions easily. Within a few years, he had purchased his own shop. His confidence, positivity, and professionalism helped him procure a major contract to make wooden boxes for the American chain Pier 1 Imports. After that, he turned to dealing, employing many "boys" to do the simple carving work. He also began selling drums, because they were more lucrative than woodcarvings. Eventually, he moved to England and opened a Ghanaian craftshop in one of London's large markets. When I first met him in 2005, he was returning to Accra several times a year to buy drums, do some carving himself, and fill shipping containers with work made by some of the master carvers who had originally taught him. In so doing, he supported a lot of talented older craftsmen who weren't getting much walk-in business. By 2010, his success made it possible to purchase a farm and build a house up north; he was also planning to develop a bespoke wood furniture business.

Dawson was a Carvers' Lane success story, someone who had parlayed carving knowledge into an international business that allowed him to reenter the Ghanaian economy as a patron and to build a house in his hometown. Still, his success was tinged with sadness. He was an artist by nature, not a businessman. For Dawson, the essence of creativity was the ability to create a space of attentive openness, to invite a new image to appear, to instantiate that image in wood, and, through the instantiation, to make new forms. Like other carvers, he considered Awoonor to epitomize this approach.

> One man that I respect a lot is Mr. Awoonor. He was a man who used to create a lot. And he is someone who doesn't all the time do spoon[s].

And he's someone who can maybe change spoon into knife, or maybe something else. He doesn't think about the money he will get today or tomorrow. If he were to think of the money he should get today or tomorrow, he will never carve those figures. You understand? Because those figures take a long time to carve. He can carve elephant, he can carve giraffe, he can sell them quicker. But he doesn't think of doing it all the time.

In this quote, Dawson articulates how neoliberal time was inimical to artistry. As the years passed, Dawson saw Awoonor change in these ways, looking up hopefully each time someone passed his shop, instead of focusing on his work. Economic insecurity took away his time and attention. It destabilized his inner vision, forcing his gaze out toward the dwindling market and financial concerns. Dawson explained, the repetitions in his speech punctuating his sadness:

It's just—the money is not there. The money is not there. Otherwise, the man would create. At first it was like, you work, and you won't find it [very] difficult to sell it. But as at now, you will find it more difficult to sell it. The system is hard now, so everyone is looking for money, not thinking about the work that he is doing. But at first, a lot of people were thinking about the work that they were doing, you know? At first, you would be in your shop, and the person would come and buy from you. But now, it's not like that again. You have to chase here and there before you get somebody to buy from you. So it's like, even if he is working, he doesn't concentrate on the work that he's doing. He is trying to get someone who will buy from him, rather.

Economic uncertainty also produced homogeneity, because craftsmen were too precariously positioned to imagine or risk making something new. Dawson saw this in the monotony of objects that carvers made again and again.

The work itself is lost in Arts Centre. It's lost. All that you see is the *same* things that we saw so many years. There's nothing different. But artwork—something to hang on the wall, something to use in your sitting room—it's supposed to be changing. But this, what our ancestors started doing ... that's what we are still doing. People will be fed up. If I buy elephant today, you don't expect to buy the same elephant tomorrow, but we keep on doing the same thing, the same thing, the same thing. You go and come back; you see the same thing. You go and come back; you see the same thing.

"You go and come back; you see the same thing." Dawson's iterations mapped the circular, enduring nature of the neoliberal present, in which people were unable to introduce a novel element that would propel them into a different future. Instead of the promise of time as progress, men saw themselves sliding backward.

"Walking Backward" in Neoliberal Time

Older carvers often nostalgically recounted earlier moments when business at the Arts Centre had been better. These included the early years, in the early 1980s, when they worked in a communal shed "under the tree" in the Centre's main plaza, so that their work was the first thing a tourist saw upon entering the Centre, and the early 1990s, when the Arts Centre was a stopping point for all Ghanaian tours. Private buses would pull into the plaza and large groups of tourists would emerge to scour the Centre for souvenirs. But by the turn of the millennium, the carvers had been moved to Carvers' Lane, tour buses no longer stopped at the Arts Centre, and tour guides took their clients elsewhere to shop for similar goods under more pleasant conditions.

Carvers blamed the falling market on a variety of interconnected factors, including the government's failure to support craftspeople on state-owned grounds, new timber industry regulations in the early 2000s that raised the cost of wood, and the entry of "those educated ones" into a trade they had formerly disparaged, using their networks to garner international contracts at trade fairs, setting up crafts shops in elite neighborhoods northwest of Central Accra, and outsourcing these orders more cheaply to carvers outside the capital.[2] For carvers, developments functioned to exclude them from the profitable international trade they had helped develop.

Other trends in global trade also affected the market for Ghanaian handicrafts, less immediately visible to carvers (Boampong 2015, Kasfir and Förster 2013, Mahoney 2017, Wolff 2004). According to Ghanaian wholesalers and foreign buyers, the "African" aesthetic had its heyday in the 1990s, to be replaced by an interest in "Indian" things in the 2000s. In addition, better pricing and quality control led to the handicraft market shifting partly to Thailand, where elephants and other "African" objects could be produced more rapidly, with greater uniformity, and sold more cheaply, than in Ghana. Carvers at the Arts Centre could do nothing to mitigate these changes. Once the dominant craft form, by the turn of the millennium carving had become a residual trade at the Arts Centre. Awoonor—standing at his station, looking up worriedly from the beautiful figure he was carving, to find no one coming to buy it—seemed to embody this change.

When I first met Ametewe in late 2003, he, along with most of the other senior carvers, was already struggling economically. He was a single father of two sons, constantly worried about bills, school fees, medical emergencies, and other expenses. His shop was almost always empty, its silence in striking contrast to neighboring jembe drummaking shops. The situation only worsened over the next decade. As life in the city became increasingly untenable, many carvers were forced to share their workshops or rent or sell them to newcomers. Each time I visited Ametewe in Accra between 2004 and 2009, the space taken up by his workshop, source of his agency and power as a man, was smaller than the last. In 2006, he had split his shop in two, putting up a wooden partition so he could rent out half of the space. He further subdivided the shop in 2009.

By 2014, he had left the Arts Centre and returned to his hometown in northern Ghana.

> In the villages, when you are in the villages, that is your home. And, you can farm. At least you can have your daily bread. But when you live in Accra, and you cannot pay your rent, you cannot pay water bill, you cannot pay electricity bill, you cannot take car and come to work, you are forced to go back to your village. There, if you take your cutlass [machete], you go to your farm, you can farm. By the end of the day, at least you can have your food to eat. You can fetch water from the rivers. You can fetch firewood from the rivers. At least, nature can sustain you in the villages. But when you live in Accra you cannot do all these things. And even, maybe even you cannot pay your children's school fees. But when you go to villages, maybe the school fee is low, you can be able to make amends and pay. There too, when you are there, originally you have a profession [carving] to do. So once in a while you will carve and come sell and you will use that one to pay your children's school fees ... You are in your home, even if it is just a thatch hut, so no one is coming to [demand you] pay your rent. So you are a bit free.

By the mid-2000s, none of the carvers were making enough money to pay basic bills, support their families, and educate their children, much less support apprentices. Ametewe's only option was to abandon the capital and the hopes and aspirations that had brought him there. Instead, he had to "walk backward" to the village, which he had temporally and spatially left behind in his quest for upward mobility for himself and his children. Awoonor's shop underwent a similar downsizing, and by 2015, he too was living back in the village. These carvers, and others like them, experienced economic change as an existential impossibility. The interconnected progression through carving toward adulthood had thus been upended by changes to Ghana's economy, replacing narratives of advancement with those of decline (Ferguson 1999).

It was not only established carvers who were affected by these neoliberal reversals. When men gave up their shops or stopped performing as masters toward apprentices, they unintentionally deserted a generation of youth, leaving them to fend for themselves. No longer able to realistically imagine the progress from sleeping in a shop to owning a house, millennial youth were left floundering in search of a new pathway to adulthood and independence. One adaptive strategy was to turn away from the slow social and artistic time of apprenticeship and fine craftwork toward new, more malleable forms of craft labor that seemed better aligned with the demands of neoliberalism.

Hustling toward Rastahood

Without the option to become an apprentice in the protective space of the craftshop, boys were forced out into the public spaces of Accra. Unaffiliated with shops, they became hustlers and went around "chasing the customer" within the

Accra touristic contact zone, searching for potential clients at the Arts Centre, airport, beaches, hotels, and state-run tourism sites. Within the Arts Centre, they steered tourists to particular shops in exchange for a "dash" or percentage of sale. As Dawson described,

> Yeah it started about six years coming now—'96, '97—at first, we didn't have boys going up and down, chasing whites, chasing those who want to buy. It was not like that, you know. ... And those who do the work actually are not there now, again. Because at first, there were a lot of workers over there. But now it has turned into something like, only dealers, in such a way that we shift into a business that makes money, not into the work that you are born to do.

Many senior men blamed the boys for scorning the labor and artistry of carving, and for driving customers away from the Arts Centre. But Dawson understood why they had stepped out of the workshop and into public space, out of making into selling, out of taking time to learn a trade to "chasing whites."

Hustling evinced the demise of the old, functional system of craftwork. It incarnated economic change as a deep loss that extended beyond a shift in labor practices into a foreclosure on the known future, "the work that you are born to do," as Dawson described it. As Jeremiah Nachinsi, a hustler and jembe seller, reflected:

> Arts Centre here is crazy. You can see that there's a lot of people walking around; you don't have anything to do. How many customers will come to satisfy everyone [so that] each and every one will earn something? And then put something in the bank, to feel like, "Oh, one day, if I become old, I can take this money to live and take care of my family or some other thing."

One October afternoon in 2004, I spoke with Jeremiah and his friends as they sat on the tree trunk in front of Akwaaba Restaurant. From this vantage point, they could see the entire central plaza connecting the Main Gate to the Arts Centre building and the entrances to the Handicraft Market, the Textile Market, and Carvers' Lane. Any potential client who wanted to visit any of these sections had to move through this public space, where they would quickly be surrounded by hustlers. But business was slow that day and no tourists were in sight.

The youth sitting on the tree trunk were in their 20s, and they were all from the northern city of Bolgatanga in the Upper West Region. Each of them had traveled south to the capital by bus in the 1990s. Samson had been in Accra the longest, since 1990, while Noah and Koz had arrived in 1997, and Moab and Jeremiah in 1999. Initially, they had made a meager living by weaving and selling the ubiquitous Bolga baskets, for which their hometown was known, in shops owned by older men. Koz explained, "We have been weaving them since we were children. That is our main work. That's what we knew before we came and learned this." The problem with basketweaving, as with carving, was that it could no longer sustain them, so they had turned to hustling and selling jembe drums instead.

I asked who the first northerner had been to come to the Arts Centre. Bearing out the punctuated, spatialized temporality of the field, Noah quickly disabused me of the assumption that this history was linear or progressive, like the traceable lineages of master carvers and apprentices. Instead, hustling was an equal playing field where free agents competed freely:

> See, this is the hustling field. It's hot, so anyone at all can come. It's not like Ministry business where you can say, "This person from the North was here." So we don't know who was here first. Some came [to Accra earlier, but] they are out of town now. Some are in Europe, making it big, but we are still here, hustling, trying to make it in life, which is really hard, which is why you see us sitting down here.

Hustling was precarious because customers were a limited good, and fair game for anyone to go after. So these close friends worked as a unit and shared their profits.

The hustle was familiar to many working in Ghana's cultural industries and informal economy (Chernoff 2003, Schauert 2015, Shipley 2015). It meant hard work, survival, being always on the lookout for potential resources, ready to maneuver to harness them. Indeed, some hustlers had made it big and were living in Europe. The ideal hustler was rarely still. He was always moving along with the tourists, talking with them, trying to find out what each one needed so he could take them to the place where they could get it. Kennedy explained the role of hustlers in the zero-sum economy:

> We have shops, but we all sell similar things, so you have to use your brain and your fastness and your kindness to get a customer to a shop, and if something attracts him or her, then he will buy from you. But if you don't make that effort, and you are sitting in your shop, for the customer to walk there is not easy, because we all sell similar things. So people go after customers.

Noah concurred, describing the vigilant waithood that characterized hustling:

> To get someone to come to your shop, you need to go out there, talk well to the customers. Because they are foreigners, they don't know anything about Ghana, so when you talk to them in a way that they are happy, they will come to your shop. Right now, we are sitting down. But we are not just sitting. We are watching out there, looking for customers, so that we can introduce ourselves .... Our dream is to expand our business to abroad; we are looking for anyone who can assist us. ... So that is why we are sitting here.

Ultimately, they all felt that a successful future hinged on connecting with tourists, not just as customers but ideally as potential business partners or patrons.

The problem was that tourists hated the hustle. To shop for tourist art was to experience the quest—the meandering exploration, the adventure of learning, the thrill of discovery, the sudden snare of an unexpected object (Price 1989). Hustling turned the hunter into the hunted: tourists felt accosted, invaded, unsure of the proper interactional script. Hustling tainted the Arts Centre's reputation so much that the 2004 tourist guidebook *Lonely Planet* described it as "best avoided." Shopowners loathed the hustlers because they repelled tourists, destabilized an already tenuous playing field, and denatured both production and sale. They vented their larger frustrations onto the youth, who, in place of respectful apprentices, now chased and bothered the tourists.

In return, hustlers resented the way senior men uncoupled youth behaviors from their own lack of support, disassociated themselves from their own pasts when they had been dependent on older men, and more generally failed to recognize how the socio-economic reality of the marketplace had changed. Enan, who had recently made the transition from hustler to shop co-owner, commented, "Life is not easy, and hustling is how I started. ... And the old men forget ... Now that they have shops, they have forgotten and are trying to prevent others from doing the same." Without the patronage of senior men, hustling was the only path toward shop ownership and independence.

Despite such justifications, many youth were ashamed to hustle. Hustling epitomized their dispossession and desperation, proved that they had neither thing nor place of their own. As Noah said sadly, "You know, sometimes, we get chased out of here, because we don't have shops. It is not easy to get a shop here, that's why you see us running up and down. And that's why you see us sitting down here. We are not lazy, it's just that there is no job for us to do."

Hustling took so much dignity out of labor that would-be compatriots sometimes became enemies. In the enduring present of neoliberal time, where every moment persists only as the crisis of waithood and survival, everyone was a competitor. Even Jeremiah, whose calm, peaceful nature was known to everyone, had gotten into a couple of fights with other hustlers at the Arts Centre who had interrupted sales he was trying to make. Describing an encounter, Jeremiah said, "All of us who are hawking, we feel like it's no man's land."

This was a no man's land devoid not of people, but of rules, where it was difficult to attract and build relationships with the kind of international clients who could help achieve the hustler's ultimate goal: to step out of the circular present and into a linear future defined by shop ownership, financial independence, and adult masculinity. Jeremiah continued,

> Maybe that customer may be the customer who helps me, gives me more orders, so that I can make some small money. ... I pray I should meet a business partner and do business so that I can do export and import and I can then go back to school. This is the time for man to struggle for life.

Hence, for youth as for men, the struggle precipitated by the closed circle of neoliberal time could be its own trap, always reproducing the competition and strife that subsumed productive activity.

Hovering in Place, Waiting for Change

While scholarly attention to millennials as the generation most shaped by neoliberal temporality is understandable, its effect on their elders and on intergenerational relations is essential. In millennial Accra, two interdependent generations of men experienced crises of labor and time with the onset of neoliberalism. Their experiences both diverged and converged; they regarded one another sometimes with empathy and sometimes with recrimination. While youth experienced precarity as a loss of support and protection from the men in their lives, leaving them to hustle for themselves, senior men, unable to pivot to new forms of work, saw their power dwindle and their sphere of agency shrink. Their delegitimization seemed to be exhibited in the unruly bodies of youth, as their former apprentices ran wild. And yet, both generations mourned the loss of ethical relations between men and boys, men and work, and work and time, and the abeyance of security that followed. The demise of the master-apprentice workshop institution generated professional and existential breaks in the life trajectories of both partners in the exchange. Eventually, imperfectly, these would be sutured by two interrelated forms of craftwork that promised to accommodate neoliberal time: hustling, which materialized the circularity of waithood and the ongoing present, and jembe drum production (discussed in the next chapter), which objectified sped-up time and precarious, flexible labor.

It is this state of "waithood," typically applied to a generational experience of remaining stuck in the transitional category of youth, which most aptly characterizes the "neither-here-nor-there state" of hustling (Honwana 2012:3, Singerman 2007, Sommers 2012). If apprenticeship was the achievement of mastery through time— a direct, forward movement from boyhood to manhood—the hustle spatialized arrested development. If the workshop was a space of possibility for working-class masculinities under capitalism, hustling in public spaces embodied waithood under neoliberalism: its stuckness, circularity, and anxiety; its repeated, failed attempts at self-generation (Ferguson 2006, Mains 2007, Piot 2010). Not having a place of your own meant not having a clear temporal trajectory that would carry you from the present into the future, from dependence to independence. The hustler hovered in place, marking time in an endless present characterized by the "indeterminacy of the possible," a tense, activated, patient waiting in a laterally oriented temporality that ignored the near future in favor of a distant utopic one (Guyer 2007). Many hustlers tolerated the precarious waithood of the present by imagining a more stable future in which they could become men. As Samson typically dreamed, "One day, one time, we hope it will get better for us, so we [can] make big money … and build our mansion houses and be with our wives and children. So we pray for long life for all the hustlers in Arts Centre." Such future imaginaries may be seen as "messianic, salvific, even magical manifestations" of capitalism in its millennial forms (Comaroff and Comaroff 2001:293).

An important characteristic of this orientation toward the future is the way it makes the subject, who is trapped in the present, entirely dependent on external forces to renew the possibilities of the subject's own agency. As Lauren Berlant has written, the neoliberal—a heuristic pointing to a set of delocalized political and economic processes—created a new historical sensorium, an intimate public of subjects who circulate scenarios of contingency and exchange paradigms for "how to live on, considering" (Berlant 2011:9). How to live on, considering that the old life scripts no longer apply, that the future promised to the subject is gone? In the protracted present of neoliberal time, nothing happens except precarity. It is thus, as Berlant describes, a state of things in which something that *might* matter is unfolding, a sense of emergence of something that *might* become an event (Berlant 2011:11).

In this time of precarity, both men and youth in Carvers' Lane sat or wandered around, waiting for something to happen, for some *event* to emerge out of the quotidian, for the arrival of some outside force to shake up the static space of the present. For millennial youth trapped in a precarious waithood, these outside forces were the introduction of the jembe drum and Rasta music to the Ghanaian art scene. From these, the hustlers went on to create a new art form—Rastahood—that would enable them to reenter a chronotope of possibility, of movement from boyhood to adulthood, from Ghana to aborokyire. Or as Noah exclaimed, "We are going to rule the world! ... Not to rule the world by gun, but by wisdom, because we believe in Jah, and we believe that Jah is going to help us be the people we want to be."

Notes

1 Seminally defined by Sahlins (1963:289–292) as a status obtained via personal achievement and power rather than inheritance, "big men" attain status, power, wealth, and renown by harnessing the productive labor of loyal, lesser men to their ambition and then exhibiting "generosity" to their followers. See Lindstrom (1981) for a history of the "big man" concept in anthropological theory. For use of the big man model to explain most non-state, kinship-based, patron-client relations in West Africa, see Barber 1981, Hyden 2006, Jackson and Rosberg 1984, Utas and Nordiska Afrikainstitutet 2012.

2 Such fairs were collaborations among governmental, non-governmental, and foreign agencies that promoted export and handled international labor regulations, quality control, and assistance and training for artisans. The fairs provided the main opportunity for craftspeople to present their wares to international clients (such as Pier 1 Imports or TJX), but individual carvers or carving collectives could only access these fairs if they had contacts with the organizational bodies that ran them.

3 From Elephants to Drums

Object, Performance, Mobility

By the early-2000s, whenever carvers looked up from their dwindling work, they saw the Arts Centre transformed. Workshops left empty by fellow carvers were newly occupied by jembe drummakers. Mahogany, rosewood, ebony, and teak figurines were replaced by goblet-shaped drums, which seemed to be everywhere, filling and spilling out of the shops, onto the sand under the Lane's trees where carvers had once stockpiled wood, and into the public arenas of the Arts Centre. The regular carving rhythms were interrupted by sporadic, explosive beats on drums being tested as they were made or sold. In a voice filled with confused wonder, Ametewe said, "It just *opened*. So most of the people who even were carvers decided not to do the carving—they jumped into jembe. … Those people who do any other work—they all jump into the jembe. So when you look at the place now, you can see that jembe dominates every aspect of Arts Centre."

Only a few years prior, this would not have been imaginable. The jembe drum may have been in the possession of various Ghanaian musicians and individuals here and there. But there was not a noticeable presence or phenomenon, institutional or informal, centered around jembe production. It had no significant presence and then suddenly it seemed to be everywhere. At the Arts Centre, those men who could pivot away from their former craftwork turned to producing drums; and youth, their former apprentices, now became their precarious employees. However, unlike carving's long apprenticeship, jembe production could be learned quickly, so with the right savvy and connections, youth could become drum producers themselves. In this way, drummaking, unlike hustling, could offer a way out of the perennial present of neoliberal time, into socioeconomic mobility. In this chapter, I trace how the jembe drum became the African icon that it is today, how it came into youths' lives, and how it transformed them. If handled correctly, the drum could transform a waiting youth into a transnationally mobile man.

Several paths had to intersect before youth at the Arts Centre could avail themselves of the jembe, since it was not a Ghanaian drum and, with the exception of cosmopolitan Ghanaian musicians who came across it while performing abroad, it was relatively unknown in Accra before the late 1990s. To reach Accra, the drum had traveled a long, circuitous, cross-Atlantic loop, in the course of which it had produced multiple intercultural performances (Myers 1995), constituting and being constituted by numerous audiences (Barber 1997:353, Myers 2002). Originating

DOI: 10.4324/9780429244568-4

in the Mande region of West Africa, the drum was resignified under French colonialism as more generically regional and subsequently became a sign of national identity in the newly formed West African states of Mali and Guinea. International tours by national troupes brought the drum to Europe and the United States, where it became a symbol of Pan-Africanism and Black pride. In the 1980s and 1990s, it became the most recognizable instrument representing "Africa" within world music, leading many Westerners to travel to West Africa to learn the drum, hear and see it performed, and to purchase it. This is how it arrived at the Arts Centre, where pre-existing infrastructures of craft making and performance, alongside economic changes, created a perfect landing place for the drum.

At independence, the Arts Centre had been a hub for the performing arts, and many drumming and dancing groups trained and performed there. Only later did handicrafts such as carving become the focus. Youth, especially those who had grown up in the Centre as children, had been members of Ghanaian drumming and dancing troupes as well as assistants or apprentices at carving workshops. By the late 1990s, coming of age when neither of these trajectories were economically feasible, they were a generation poised to take up the jembe, having both the capacity and need to do so. On its journey, the jembe had become a global symbol of "Africa" and thus, when taken up by youths, allowed them to expand their geography far beyond Accra, thus creating a chronotope of mobility and global travel, as opposed to immobility and waithood. "Africa" is associated with music in the global imaginary, with rhythm more than any other musical element, and arguably, with drumming above all (see Ebron 2002). And when most people outside of the continent think of an "African drum," they likely see a mental image of a jembe drum.

The Jembe Drum: Origins and Travels

The jembe is a goblet or "pestle"-shaped hand drum, "whose bowl and pipe combination along with the high-pitched head produce the jembe's unique sound" (Kalani 2002:7). Its three main components are the shell, drumhead, and rope. The shell typically measures 11–15 inches diameter at the bowl and 6–8 inches at the foot. The head is attached to the shell with three metal rings and a series of tension cords that create a "Mali-weave" pattern. The shells are carved by professional carvers out of a single log. The drumheads are typically of goatskin (Kalani 2002:7).[1] The drum can be played sitting or standing, the latter of which allows for a greater range of motion and physicality.

The drum hails from the Mande region of West Africa, a cultural area that formerly comprised the Mali Empire (c. 13th–15th centuries), which today encompasses parts of Guinea, Mali, the Ivory Coast, Burkina Faso, Senegal, and Gambia and centers the Guinea-Mali border (Charry 2000). Traditionally, jembes were made by members of blacksmith castes (*numu*) and playing was a lower status activity than the praise-singing of the *jeli*, a hereditary caste of musicians who functioned as geneologists, historians, diplomats, and spokespersons (Hale 1998, Knight 1984). In the contemporary Mande context, the drum forms part of both rural and urban traditions, including life cycle rites, ceremonial mask dancing, and celebrations (Polak 2000:10–11).

West African drumming, including jembe rhythms as traditionally performed, include dance and song and are defined by interaction, from the polyrhythmic interdependence of drum parts to the call and response form to interactions between the lead singer and chorus, to the ensemble's spontaneous responses to statements made by drummers (Dor 2014:146–148).[2] Similarly, the dance which accompanies drumming is a nonverbal, embodied, aesthetic means of communication. Finally, "jembe performance is participatory; people step out of the role and space of audience into that of performance and step back again" (Polak 2007:11). As a performance genre, drumming and dancing's interactive interdependence allows participants to embody "communal creativity" (Dor 2004:147).

William Ponty and French Colonial Africa

To reach its massive global popularity, the drum underwent a series of recontextualizations like those of the craft objects described in Chapter 2, resignified in colonial, nationalist, and finally international contexts. In the 1930s, the French integrated jembe into an emerging genre of African theater at a school in Sébikotane (near Dakar in Senegal) called the École Normale Supérieure William Ponty (Charry 2000, Cohen 2012, Polak 2000). Like the British Achimota College in the Gold Coast, Ponty brought together the best and brightest of France's vast, ethnically diverse West African colonies. Like the British, French authorities feared the formation of a "culturally unmoored (and thus potentially dangerous)" African elite alienated from both colonial and indigenous life (Cohen 2012:20). They hoped to prevent the development of such potential agitators by educating a group of civil servants and teachers to mediate between the colonial regime and its colonized populations (Sabatier 1977). They pinned their hopes on an educational model which, like the adapted education program at Achimota, grafted reconstructed African elements onto a French curriculum, here largely as extracurricular activities. These included obligatory summer projects in which students researched their cultures of origin and wrote plays that appropriated indigenous content (Cohen 2012; Polak 2000). In the 1950s, the French also established cultural centers from which they ran a hierarchical system of local, regional, and territorial competitions in African theater (Cutter 1971; Hopkins 1965; Polak 2000). Decontextualized, secularized Maninka jembe rhythms and dances were centerpieces in this emergent genre, institutionalized as a performance of supraethnic Africanness and transmuted into an ambivalent object of colonial power.

Les Ballets Africains, Independence and National Culture

At independence (Guinea in 1958, and Mali and Senegal in 1960), Francophone elites (many of whom had graduated from Ponty) transposed the colonial model onto the development of national cultures. In Mali and Guinea, the two new states in the Mande heartland, state-run national music and dance ensembles known as *Les Ballets Africains* recontextualized cultural forms, recruiting performers from the rural hinterlands to stage each nation to itself and the world.

Fodéba Keita, a Maninka graduate of Ponty, was instrumental in this shift and a crucial agent in propelling African drumming onto the world stage. His musical interest sparked by a family friend who had performed as a jeli at the 1935 Paris World Fair, at Ponty, Keita played the banjo and put together a band that merged European, Caribbean, and African influences. While studying law in Paris, he founded the original *Les Ballets Africains*, "the world's first professional and internationally touring African performance company" in 1952, based on a prior group he formed in 1949 (Charry 2018, Cohen 2012:12, 19,). Between 1954 and 1958, the ensemble toured a total of 170 European cities as well as French West Africa, South America, Eastern Europe, and Israel, receiving extensive media coverage including appearances in film, television and radio broadcasts, and musical albums (Cohen 2012:13).

Keita's Ballets were Pan-African and modernist, combining indigenous traditional forms, Ponty-type theater and poetry, and Caribbean music and dance forms (Cohen 2012:24–25), and thus easily convertible to national culture. In 1958, when Guinea, following Ghana, became the second sub-Saharan state to gain independence, President Ahmed Sékou Touré nationalized Keita's Ballets. Other newly independent Francophone states followed suit. Like the Ghana Dance Ensemble and Nkrumah's national theater movement, the Ballets recruited performers from different regions to publicly represent the nation through folklore (both internally to its citizenry and externally to the world public) and were thus crucial to nation-building (Polak 2000: 10–11). The nationalist context transformed drumming and dancing forms and their transmission, as younger generations of drummers and dancers were taught in urban rather than village settings, learning rhythms in a context of national spectacle rather than as part of a more localized, holistic way of life (Melnikoff, in Dor 2014:143).

The "nationalization of formerly local culture" would turn out to be an important step in the shift to liberal (postcolonial) and eventually neoliberal (postnational) capital production and movement (Polak 2000:17). As ethnomusicologist Rainer Polak argues, early African nationalism, defined by a desire for international respect, isolated and pre-arranged cultural forms into national culture. Ironically, this isolation ended up benefiting the commodification of these forms within a global system ever in demand of new forms to commercialize.

Coming to America

Everywhere Keita's Ballets performed, they sparked local interest in the jembe, but nowhere was this more influential than in the United States. By the time Keita's ensemble performed in the United States in 1959, the American stage had been primed for their arrival. From the late 1920s through the 1950s (and beyond), Black musicians, dancers, dance teachers, and choreographers had transformed African drumming and dancing into a deeply politicized sign of African-American identity

(Charry 2000, Cohen 2012). Centered in New York, the cohort included Nigerian-born Efiom Odok, Sierra Leonean Asadata Dafora Horton, American Ismay Andrews, and most critically, the American ethnographer-choreographers Pearl Primus and Mary Katherine Dunham. Dafora, Primus, and Dunham (who founded the subfield of dance anthropology) laid the foundation for staged African drumming and dancing in the United States.[3]

Then, in 1954, the Nigerian Yoruba drummer Michael Babatunde Olatunji arrived in New York at a pivotal moment, his arrival coinciding with the American civil rights movement and independence movements on the continent. "[R]emembered as perhaps the greatest cultural vanguard of the Pan-African movement" (Dor 2014:37), Olatunji was pivotal in the global resignification of African drumming. Graduating from Morehouse College in 1954 and unable to procure funding for graduate school, he moved to New York and there took up his calling as a cultural ambassador, partly funded and supported by his close friend Kwame Nkrumah (Charry 2005:3). In 1958, he founded a drumming and dancing group with African-American and Afro-Carribean personnel, as there were not many African musicians in the city at that time. A nightly performance titled "African Fantasy" at Radio City Music Hall gained him a coveted Columbia record contract in 1959, which produced six albums over the next six years.

The first and most famous of these, *Drums of Passion*, "introduced hundreds of thousands and perhaps millions of Americans to African drumming" (Charry 2005:4) and "established Olatunji as Africa's most famous drummer" (2005:4). The album was its own amalgam: with songs and rhythms from Nigeria performed in hybrid combinations of drums (many from Ghana). Charry describes this album as a compelling "African-based music composed and arranged for an American audience" that "presented Africa as a living creative force" (2005:7). Throughout the 1960s, his work was aligned with the celebration of African culture that defined Pan-Africanism, African independence, Black Power, and African-American civil rights. It culminated in the fulfillment of his dream, the opening of the Olatunji Cultural Center in Harlem in 1967.

Olatunji worked closely with many leading jazz musicians and African drummers, including one of Keita's star soloists, Ladji Camara, who arrived in New York in the early 1960s. A brilliant *jembefola* (jembe master drummer), Camara introduced jembe-based traditions to America and became the acknowledged "father of the jembe movement" (Charry 2005:10).[4] As Charry explains, "The highly virtuosic solo jembe drumming and dancing are different from what is practiced in the anglophone countries of Nigeria, Ghana, and Liberia, which had received earlier exposure in the United States" (Charry 2005:13) and brought attention to the formerly underrepresented Francophone countries of Guinea and Mali. The drum's capacity for virtuosity, drama, and solo play caused the predominantly African-American drummers who followed Olatunji and Camara in the 1960s and 1970s to create ensembles focused on the drum (2005:13).

While the jembe was being popularized in American performance spaces, Ghanaian (primarily Ewe) drumming was introduced to American universities through the arrival of Ghanaian musicians trained in the state-run Ghana Dance Ensemble

(Ghana's version of *Les Ballets*) and elsewhere in Kwame Nkrumah's state-run cultural complex (see Chapter 1).[5] Thus, by the 1980s, West African drumming in the United States had established along several lines: jembe drumming styles dominated in performance spaces and primarily Ghanaian (and mostly Ewe) drumming dominated in universities.

By the time Keita's and other national ensembles disbanded, they had established an elite class of cosmopolitan jembe drummers from different countries. These *jembefolau* [master jembe drummers] had traveled the world, made numerous foreign contacts, and were experts in presenting the drum to international audiences. Many of them settled in Europe and North America, where they became charismatic embodiments of the African continent while performing and teaching the instrument (Berliner 2005, Castaldi 2006, Chernoff 1981, Flaig 2010, Polak 2007, Sunkett 1993). Through them, the West African jembe achieved "world status as a percussion instrument, rivaled in popularity perhaps only by the conga [from Cuba] and steel pan [from Trinidad]" (Charry 1996:66). The jembe eventually became equated with Africa itself. Beginning in the 1960s, the jembe was a means for West African men to travel, establish themselves abroad, make contacts, and internationalize their performance practice. They had become, in the words of ethnomusicologist Pascal Gaudette, "jembe heros" (2013). This mobility was precisely what the youths in this book aspired to do in the 1990s, building on the reputation the jembe had garnered over previous decades while adding in their other skills.

World Music and the Drum Circle Movement

Through the travels described above, jembe rhythms became a category of "world music." First circulated by academics in the 1960s to celebrate and study global musical diversity and to include "the rest" within musicology programs focused on "the West", the term "world music" gained currency in 1987 as a marketing category coined within the music industry (Feld 2000:146–149, Taylor 1997:2).[6] Marketing "danceable ethnicity and exotic alterity on the world music and commodity map" (Feld 2000:151), world music constructed an "experience of global communications and authenticity through symbolic means whose very difference depends so vitally on their sameness as transnational commodities" (Erlmann 1996:481). The category reframed and equalized a wide, diverse range of musical genres through a mix of discourses about "others." Merging discourses of authenticity, primitivism, new ageism, and universalism, the genre decontextualized, deritualized, and evacuated musics of local meanings to create a new, naturalized, universalized musical form that banalized difference (Guilbault 1993:40). It defined music as universally accessible "feeling" and erased imbalances of privilege, power, access, and gain by positing music as a happy, multicultural space of flow, sharing, and choice (Feld 1995, 2000, Taylor 1997). Much, though not all, "world music" consumption shared touristic impulses described in this book, similarly depending on resilient colonial tropes of exploration, discovery, contact, and authenticity ("fresh but timeless" Taylor 1997:28) filtered through a New Age lens for late 20th century. Such consumers fetishized a musical "other" as a portal to "the timeless, the ancient, the primal, the pure, the chthonic; that is what they want to buy, since their own world is often conceived as ephemeral, new, artificial, and corrupt" (Taylor 1997:26).

"I'm the drum, You're the drum, We're the drum." The quote above is by Babatunde Olatunji (Olatunji, Atkinson, and Akiwowo 2005:1), who in the 1980s again played an immense role in the global spread of African drumming. But while his activities in the 1960s had introduced the West to the beauty of African music and culture and to celebrate a global Blackness, his work in the 1980s as part of the American drum circle movement, which radically transformed African drumming for a mostly White audience, may have caused him a far greater ambivalence. Drum circles were vital in Westerners' exposure to African drumming, and to the jembe in particular (Carter-Ényì, Carter-Ényì, and Hylton 2019, Kalani 2004).

A key moment of signification which directly shapes the Rasta story, a drum circle, a now familiar global phenomenon, consists of a group of people sitting in a circle and drumming together. There is no hierarchy within the group, no master drummers. Begun in the 1960s as part of the American counterculture, the movement really took hold in the 1980s. According to percussionist Arthur Hull, considered by many to have conceived of the modern drum circle, drum circles are "about facilitating community through rhythm-based events" (Hull 1998:12). This community is defined by communion and connection "to the Earth our Mother, to each other, and to our collective spirits" (Hull 1998:9). As defined by Mickey Hart, musicologist, Grateful Dead drummer, and one of the drum circle movement's key figures, "the Drum Circle is a huge jam session. The ultimate goal is not precise rhythmic articulation or perfection of patterned structure, but the ability to entrain and reach the state of a group mind. It is built on cooperation in the groove, but with little reference to any classic styles. So this is a work in constant progress, a phenomenon of the new rhythm culture emerging here in the West" (www.remo.com/drumcircles).

Olatunji was crucial in this new development in African drumming in the United States. After a decade of diminished opportunities, he traveled to California in the 1980s, where he reconnected with Hull and Hart (of the band the *Grateful Dead*). These two musicians were already developing drum circle philosophies, and Olatunji found a new role as elder for a new audience (primarily young Americans following the *Grateful Dead*), for whom he developed a new drumming doctrine. In his foreword to Hull's first book on drum circles (1998), Olatunji writes of "the evocative power of the drum, and how to use it as a necessary tool of communication empowerment, bringing people together, sharing very positive vibrations amongst all participants regardless of their color, creed, religion, or gender" (Olatunji 1998:7). This American resignification of drumming, which rested on "notions of peace, love, spiritual enlightenment, and social action, found resonance in the 1980s and 1990s among white youth and adults searching for alternative means of expression and spiritual fulfillment" (Charry 2005:15). It built from perceptions that African drumming produced community, communion, and connection and as a form of public healing.

In his dedication of that same book to Olatunji, Hull explained: "[Olatunji] has been birthing 'rhythmaculture' into the biggest cultural mixing bowl on the planet, the USA. 'Rhythmaculture' is something that America lost somewhere along the way on its path to "progress, our most important product" (Hull 1998:9). Thus, despite its universalism, African drumming emerges from a familiar split between Africa and the West, wherein Africa signifies a prior, essential connectivity between all things, the West signifies the loss of this connectivity "somewhere along the way on its path to 'progress,'" and drumming is the way for Westerners to get back what they have lost. This narrative of loss (and refound gain) often attends the oboroni "discovery" of African music.

While the need for forms of public healing and being in community are real, the drum circle fulfilled the needs of an audience "less interested in the cultural background of African drumming and more attracted to the euphoria of a group experience" (Charry 2005:14). Happening concurrently with the rise of world music, the drum circle movement shared with that broader phenomenon processes of decontextualization, commodification, and exoticization. It erased the specificity of different drumming traditions and removed many of their stringent rules for entry and participation, instead rendering African drumming as something accessible to all. The drum circle was far removed from the cultural origins, norms, roles, and training that defined West African drumming, including those determining who could drum (due to cultural prohibitions around gender, age, caste, etc. and lack of training and skill). To become a drummer, one had to take part in a challenging, years-long mode of formalized training similar to the apprenticeship model described in Chapter 2. By contrast, the drum circle posited African drumming's democratic universalism, its fundamental accessibility and welcome to all who wish to participate, "regardless of their color, creed, religion, or gender" (Olatunji 1998:7) and also, regardless of skill. As Charry explains, it was "a universal philosophy of peace and love, rooted in Africa yet malleable enough to be shaped by Americans for their own needs" (Charry 17). Drumming provided an alternative means of expression and spiritual fulfillment of "American imaginings of a global utopia" (Charry 17).[7] Drum circles transformed and simplified African drumming for a Western, US-centered context, replacing the polyrhythmic relation between complex drum parts with a unified, simple rhythm played by everyone. This was essential to their function as utopian, all-inclusive sites of healing, spirituality, and universal consciousness through shared rhythm. As Hart explains, "The ultimate goal is not precise rhythmic articulation or perfection of patterned structure, but the ability to entrain and reach the state of a group mind" (2004).

As anthropologist Paulla Ebron reminds us, "we have learned to imagine regions through repetitive tropes" (2002:10) and Africa is "the geopolitical ur-site of performance from the perspective of the West" (2002:16). More specifically, Africa is rhythm and community feeling (2002:33–34). Africa is the drum. As Ebron points out, these features "become facts not just about 'music' but about … an idyllic unstratified Africa, the Africa of 'African music'" (2002:34). This "positive" representation posits Africa as opposite to the West, which emerges as "individuated, distinctive, and complex" (34). Music becomes emotion to the West's reason, collective versus individual, integrated into a holistic social life as opposed to autonomous within a society defined by alienation.[8] This split means that people of European descent, "Assured of the capacity (as a group of people) for rational thought and reason, [...] can enter into African drumming from a position of privilege" (Charry 2005:19) while simultaneously experiencing a profound lack. And African drumming is envisioned as a "communal property whose spiritual qualities are shared and experienced by all; in short, it is an art form that can and must communicate with people of all races and cultures" (Bebey 1975:vi).[9]

Aspects of African drumming facilitated its transfer to this new context and use. The interactivity that defines it allowed people to generate "a better understanding of mutual co-dependence, egalitarianism, or the strength of collectivism" (Dor 2014:147). It was ideal for creating an ecstatic experience of *communitas* (Turner 1969). Add to this the virtuosity, physical skill, infectious joy, and beauty of the performers, their costumes, and their instruments, and it was clear why it was loved by aborofo. However, other elements of West African drumming challenged its easy uptake by aborofo: the contextual rules that determined who could and could not drum, the complex relationships between drum parts, and the years of training required to play well were often lost in the transfer. Instead, the jembe in intercultural settings produced an experience of cosmopolitan connectivity and intercultural communitas, a world temporarily stripped of structural hierarchies and inequalities (see Turner 1969), in which music became a means to enter and experience other cultures, a mode of "world-mindedness and world-consciousness" (Dor 2014:139). What youths were actually making by playing the drum was a new, commodified version of a ritual experience of Africa with a long history across the Atlantic world. This desire was of course gendered, affecting men and women differently (see Chapter 5).

The Jembe "Returns": Local Transformations

In its countries of origin, global forces transformed jembe drumming. According to several jembefolau and jembe scholars, jembe was traditionally a low-status, stigmatized activity that stood in contrast to the privileged status of *jeli*, or praise singers activity (Camara 1992, Charry 2000, Knight 1984, Gaudette 2013:299). But the resignification of the drum in the West created a new, alternative economy of status, capital, and mobility at home, exemplified by the careers of the jembe drummers described above. In his work with young Guinean jembe drummers in 2010s and 20s, Gaudette describes a culture parallel to Rastahood, in which youth seek to follow in the footsteps of their elders, who have successfully built global networks and trade routes for jembe music (2013:295). Gaudette traces "the jembefola's path," a pattern of out-migration in search of wealth and opportunity, followed by a return like the Mande heroes of old, who were exiled from their villages, accrued wealth and power in foreign lands, and came home in triumph, transferring foreign wealth to the village (2013:297–298; Bird and Kendall 1980, see Rouch 1967, Stoller 2002). In Ghana, when youths sought to exit a system in which they were stuck in perpetual waithood by taking up Rastahood, they too hoped for a heroic return and a rooted cosmopolitanism (Appiah 2006). The ideal Rasta life, as we shall see in this chapter, was a cosmopolitan one in which Ghana and aborokyire were bridged, with homes on two continents and a business that linked them. Youths and jembe heroes were two examples of a much larger pattern of young men seeking solutions to the sociocultural dislocations that neoliberalism had created in West Africa. For many of them, the best solution was to leave their home countries, and their only way to do so was by using their bodies and culture as resources.

Hence, we can see how from French colonial Africa in the 1930s and Les Ballets Africains, through the introduction of African drumming to the United States in the 1950s and 1960s, to the global phenomena of world music and the drumming circle, the jembe has traveled a route shared by many African cultural artifacts, including Ghanaian craft objects like woodcarvings or masks. Tracing the history of the jembe's transformation from Mande instrument to global icon, the parallels between its journey and that of Ghanaian craft objects like woodcarvings or masks make sense. Like them, it had emerged in local, often ritual, contexts and been redefined in the context of colonial education. As African states approached independence, elites politicized by Pan-Africanism, anticolonialism, and protonationalism had made the drum a sign of nascent national identity, thus poising it for its circulation within the Black Atlantic and finally for its commodification under globalization. Francophone economies experienced the same liberalization and opening up to media and tourism that Ghana did, allowing jembefolau greater contact and mobility. Drums, elephants, and masks similarly blended indigenous traditional forms, Pan-Africanist nationalism leanings, and European ideas about Africa. By the 1980s, the drum triple-voiced a colonial exoticism, a local authenticity, and a universal counterculture.

The Jembe's Journey into the Accra Arts Centre

By the later 1990s, the ripple effect of the jembe's transformations reached the Arts Centre, where they collided with the pressures placed on craftspeople. As Ametewe says at the beginning of this chapter, all of a sudden, the drum seemed to be everywhere, when before, it had had no visible presence. Drums in general were not produced in Carvers' Lane before the jembe's arrival. According to all the youths I interviewed, the first jembe was brought to the Arts Centre by a traveling Ga highlife drummer named Ajete in 1997. Ajete learned to make jembe in Senegal (or Guinea or the Gambia, depending on whom I asked) and then returned to Accra to produce and sell them. To advertise his wares, he put a jembe atop a high stick so it could be seen from "the top," the main entrance of the Arts Centre.

By the time the jembe arrived in Accra, many of the youth at the Arts Centre recognized the significance of its global circulation compared to Ghanaian drumming traditions. Gideon, for example, noted, "I was 17 at that time [in 1997] that jembe was introduced as the best instrument in Ghana ... We all tried to go through that, because jembe was selling well. All the people coming from all over the world, they all want jembe." This was the local view of the drum's transformation into a global icon of the continent.

"At that time, there was no international drum like jembe [at Arts Centre]. They only had *kpanlogo* and the other sets of instruments from Ghana," Gideon Namoale, one of the first youths to learn the drum, recalled.

Jembe was played mostly in French countries, and there are many more French than English [speaking] countries in Africa. So *jembe* was going through Europe, USA, Asia, in different parts of the world, you know. Ghana is very small and our instrument—*kpanlogo*—can't go far, because it's only from Ghana. But this jembe came from all over Africa, so we tried to learn the jembe too.

In this quote, Gideon addresses the jembe's international journey and the way this journey had exposed people around the world to the drum. As someone who grew up in Accra, Gideon centers the Ga (Accra) *kpanlogo* drum set as "our instrument" in a national sense, but he extends his statement to include other Ghanaian drum sets when he makes a distinction between the "international" nature of the jembe and the fact that Ghanaian instruments "can't go far." He ascribes this internationalism to a greater preponderance of Francophone states in Africa, perhaps obliquely referring to the French media infrastructure that enabled the global flow of Francophone music media, and for which there was no British equivalent.

The unfamiliar drum, with its unique pitch and tone and its dramatic mode of play, fascinated a lot of the youths who had grown up in the old officers' barracks near the administration building at the Arts Centre. One of these was Shem, the son of a woman who ran the chop bar frequented by drummers and dancers.

Before leaving to join a Ghanaian performance group based in Europe, Ajete taught Shem how to make jembes. Shem began making and selling drums to tourists outside his mother's restaurant, where they soon proved enormously popular. He immediately began to teach other boys. He was soon joined by Samson, on his hawker's route. He was drawn to the drum as soon as he saw Shem making it. "I saw Shem, and I saw myself doing it too. ... I didn't ask myself that I'm going to do it. I just do it." As the phenomenon grew, the hawkers, hustlers, assistants, and apprentices all shifted to jembe. Some were still in school at the time. They began building drums after school and using the money to pay for their school fees. Others, Gideon explained, had graduated or dropped out of school and were "sitting in the house doing nothing." The boys spent their afternoons at the Public Works Department (PWD) barracks, learning from each other how to make (and play) drums and then hustling, "moving together looking for a customer."

Gideon corroborated this instant attraction to the drum, this sense that it had miraculously appeared and opened up a new way forward:

[When] I started making these vibes, it is just because I am near the people who are making it, so each time I was watching and learning it. I was doing it, but I do not know exactly what I'm doing, because I do not have too much meditation on it. After, I see it is something good because it's a forever and forever creation, not only for me, or not only the rest of the [youth]: more people will trace what we are doing and it can be *big*, it can be work for many people.

In this quote, Gideon brings together many of the points made in this chapter. First, he explains how youth at the Arts Centre acquired craft knowledge informally, "just because I was near people who were doing it," almost unaware of the significance of what they were learning until later. Eventually, the jembe's value became clear. As Gideon explains, the drum could be *big*—it could generate earning power for many people for a long period of time, because of its appeal to people from "all over the world," and the youth at the Arts Centre were perfectly situated to capitalize on its allure.

From Elephants to Drums

Neoliberal pressures experienced throughout the Ghanaian economy manifested in both production and performance, and the transformations in work marked by the arrival of the jembe objectified these changes. Jembe work was "cheap," in the words of Ametewe. It was quickly learned and quickly done. Youth had begun to acquire some of the skills of the old crafts, but these were no longer economically viable for them. Jembe craftwork was fast and flexible: it not only fit neoliberal time but also replicated its precarity and short futurity. Coming of age at the Arts Centre, a site designed for the performance of culture and the sale of tourist art, the boys had access to everything needed to run a craft business. As Samson explained, echoing Gideon's description of knowledge acquisition, "When you are growing up at Arts Centre, everyone can tell you to do things: 'fetch this,' 'do that,' call him.' And you do it, and it is only later that you realize you have learned a skill. At first it is just something you do." Through situated learning in a community of practice (Lave & Wenger 1991), the boys learned how to run an international business. Connecting and communicating with tourists, procuring materials, hiring and managing freelance workers, loading shipping containers, filing port paperwork, handling international money transfers—all this and more had been done at the Arts Centre for decades.

They transferred their old skills to the new craft. But in contrast to the slow, clearly ordered transmission of knowledge "under a disciplinary society," as Ametewe had termed it, now youth informally and quickly learned from other youth. There was no age hierarchy of teachers and learners, masters and apprentices, in jembe production. Instead, the boys "moved together," as they did when hustling. Drummaking thus sidestepped the age-hierarchized system of apprenticeship that traditionally shaped craft knowledge transmission and gave elders power, respect, authority, and control over youth's labor. Master carver Ametewe provided the following comparison between carving and jembe-making, spelling out the elders' opinion that youth chose jembe production in an agentive refusal of apprenticeship, a disinclination to undergo the slow discipline required to learn to carve:

> You see, the younger generation nowadays, they don't want to live under any instruction. … They want to live on their own. And jembe-making is one of the cheapest things you can acquire. Within one day you can be able

to master how to produce jembe. So ... they choose the easier way around. That's why when you see them, as young as they are, they classify themselves as masters. But in the matter of fact, they are not masters, because they don't have shops, they don't have any capital, nobody will see them and give them orders and ask them to do any work for them. ... You know, eh, jembe work is very cheap, provided you have the capital. You go to the bush, you buy the frames, that is the only work you will do. There are people who do the carving, there are people who do sandpapering, there are people who will do the—how do they call it—how to put the skin on it. *Ehe.* So, it's there the work becomes a division of labor. This man does this piece, the other man does this piece, the other man does this piece. ... Not like carving, whereby you have to start at the beginning and do the polishing [at the end]. To become a carver is no small thing. So a lot of young people, you see, they don't want to squeeze themselves so much. They are lazy somehow. So they want the easiest way to life. So they decided to jump into jembe-making.[10]

Ametewe's phrase, "very cheap, providing you have the capital," points to a non-monterary conception of value as measured in time. In this apt, if negatively valued, description of jembe production, Ametewe names all the reasons why young men chose it over other forms of craftwork. And indeed, many of the youth hustling at the Arts Centre soon turned to jembe-making.

As Ametewe states, jembe production at the Arts Centre (and elsewhere in Accra) was broken up into several steps, and the owner/seller of the drum could outsource the work rather than doing it himself, focusing instead on procuring tools and materials, including shells, goatskins, steel rings, rope, and wood polish, overseeing production, and handling finances, communications, contracts, and logistics. It was a fragmented, uncertain, entrepreneurial activity, transforming the individual into a neoliberal subject of production and precarity.

The first, essential, and most difficult step was the carving of the shell, or frame, and this was done by professional carvers in several villages outside of Accra. The rest of the work was done by youth with little training. They sanded the drum shells, taxing but unskilled physical work. They decorated each shell with simple relief designs, beading, paint, or metalwork. They wove the rope around the metal rings. They wet, mounted, and shaved the goatskin drumhead. The skins came from Northern Ghana or Burkina Faso and were brought to Accra by leather merchants. Originally purchased in Fetish Lane at Timber Market, not far from the Arts Centre, by 2004 leather merchants brought these skins directly to the Arts Centre, as jembe production had created a high enough demand. Rings were commissioned from two welders working at the Arts Centre, again pointing to the amount of business present on the premises. Finally, the drum was tuned.

Younger than the other carvers, Afrane Dawson had more empathy; he remembered being an unmoored boy himself. When he began winning large handicraft contracts with foreign buyers, he took on young hustlers as apprentices in woodcarving, and then saw them transition to making drums. For him too, the sounds of drums filling the Arts Centre were signifiers of loss.

> I would say about 60-70% of the boys over there [making drums] were my boys. And a lot of them did not complete what they were learning. And you know, artwork, especially carving, if it is not given to you, you find it difficult to do it. That's why you see a lot them now doing drums. Because it's only drums they can [make to] survive. They cannot do any other thing. And drums ... I in particular do not call it woodcarving. You just ship it and do a little design, just put the leather and pull, that kind of thing. So if you pick up somebody from even out of Arts Centre, if you bring him there, teach him for a month, you see him doing it. ... So a lot of boys that I taught, they are boys who design this thing now. They are not sitting down doing something that will be different. No, they would prefer to design some drums and make money.

Perhaps surprisingly, youth were no less ambivalent about jembe production than their elders. Youth understood the fragmented nature of their labor. They too wished for a future or imagined a past when more holistic forms of work and personhood were possible. Certainly, they assumed jembe production was only a temporary step along the way to adulthood, rather than a life-long career identity. They viewed drummaking as a response to the lack of support that should have come from the adult men in their lives. As Noah asked rhetorically,

> Big men—what do they want you to do? There is no work outside. And if you have been learning in a shop for four, five years and you have a fight with the master and he kicks you out—where are you supposed to go? Things are worse now than they were before. They are not even giving jobs to the unemployed. ... Here they are fighting small boys when they should be fighting with all these outside people .... You know, a lot of people think those who make drums are uneducated. Some of us here, we are all educated, even though we do not have that much level of education. Because of poverty, we can't make it up, and we drop out. That does not mean that we do not have brains and we can't challenge big man's sons or big man's daughter. But that is the thing. We do not have money to continue our education. So that is why we branch to the business side. So we can live day to day. So we can survive.

The drum objectified the destabilizing effects of neoliberal space-time. Everyone agreed on the slowness of carving and the quickness of jembe production. But where youth saw speed as an urgently needed advantage, elders saw it as a direct threat to their way of life. If hustling, described in Chapter 2, represented the demise of a carving labor model and materialized the circularity of waithood, jembe production was carving's replacement, a quintessential way of working that objectified sped-up time and precarious, flexible labor. Not only was it broken down into small, repetitive tasks, the contract work required flexible, instantly available workers willing to work for little money, with no protections or benefits, for intense but brief periods of time. Boys did not know from one day to the next if work would be forthcoming, since it depended on obtaining a contract or knowing someone who had obtained a contract to fulfill in a short amount of time.

Living day by day was about all that jembe-making enabled. Life was still defined by precarity, a hand-to-mouth existence that demanded immediate returns. Only a windfall of capital made the transition from worker to shopowner possible. However, as a musical instrument, the jembe enabled youth to create a subculture—Rastahood—around drumming as performance, which gave them the potential to lift themselves out of the present, creating a new futurity. By shifting the product of labor from making drums to performing a resonant figure of Africanicity—the Rasta drummer—youth could, if all went well, achieve mobility, for the Rasta figure generated connectivity to foreign capital. Youth continually sought contracts (clients) and contacts (patrons, collaborators, and girlfriends) with aborofo. An oboroni might become a client, contracting for large orders of jembes, or a girlfriend, willing to loan startup money. Such connections, and the unexpected, though always-hoped-for and badly needed, windfalls of cash, could become a down payment on a drum shop. This is how Gideon was able to open Black Star, the first youth-run jembe shop in Carvers' Lane in 1999. An oboroni girlfriend gifted Gideon the money to rent a shop from a cane weaver who was joining the mass exodus of craftsmen leaving the Arts Centre.

At first, Gideon ran Black Star as a collective. Raised at the Arts Centre since 1980, he had seen the damage wrought by competition and redundancy. He saw in jembe a technology that could unite craftsmen and give them more power in relation to clients and the market. "[F]or me, my technology is not for we to have more shops, but we have to come together ... So when we have a customer come, the customer knows we have one big company with a whole lot of things."

But everyone wanted to be the master of his own shop (and thus, of his own self). And in the first several years after 1999, the seemingly endless demand for the drum caused everyone, as Ametewe said, to "jump into jembe." By late 2003, the original Black Star crew had split up into eight different shops, replicating the proliferation that had caused senior craftsmen to be unable to support youth as apprentices. But the association between working for oneself and successfully attaining adulthood was strong, as was the necessity to depend on oneself rather than others for care, reflecting a broader shift toward entrepreneurship in Ghana in the 1990s and early 2000s (see Schauert 2015, Shipley 2013).[11] As Moses explained the link between mastery of a shop, economic power, and mastery of self:

[E]ach and everyone wants to have their own shop. ... Because if you're not an owner, you didn't really grow up, you didn't really make it. If I'm an owner, I know I have succeeded. Also, their masters who taught them didn't give them much money to survive, they only gave them pocket money: "Take this, buy food." So most of them started thinking, "Why should I work for you?" And they saved their money so that they could work for themselves.

From Traditional Ghanaian Performance Practice to Jembe

Still, the crew continued to collaborate on the other aspect of drum culture: performance. As Gideon said, "To sell the drum, you must first play the drum." Jembe piecework alone rarely enabled youth to pierce the space of their present circumstances, but the drum carried the potential to push forward and upward into full masculine adulthood, to chart paths toward alternate futures in Ghana and abroad. They developed a subculture of performance around the drum, embodying a resonant figure of Africanicity: the Rasta drummer, and more or less simultaneously, the social world that emerges around this persona, Rastaworld. Where the jembe itself was concerned, the Arts Centre provided, at least for some youth, a historical background and transferable training for the performance arts, as well as examples of internationalized mobility and entrepreneurship to which they could aspire.

The 1950s–1960s: Independence and the Making of a National Performance Genre

As described in earlier chapters, before the Centre became a nexus for tourist arts and crafts in the 1970s, it was seminal in the creation of a new national culture with a focus on performance. This was an elite project run through the Arts Council, founded by Kwame Nkrumah, first as an Interim Committee in 1955 to plan the independence celebrations in 1957, and then later made the government's official cultural body in 1958. Like the creators of the Ballets Africains in French colonial Africa, the Arts Council's aim was to encourage the development of a national theater movement "which would at once reflect the traditional culture of Ghana and yet develop it into a living force firmly rooted in and acceptable to the new generation in Ghana" (Antubam 1963: 210). This culture was to be directed both internally, toward the new national public it sought to create, and externally, as a global representation of the new nation-state. Its aim was to constitute an Afro-modernity, rooted in and legitimized by long-standing tradition.

Beginning in the late 1950s, the Centre housed various professional and amateur drumming and dancing ensembles.[12] As described by ethnomusicologist George Worlasi Kwasi Dor (2014) in his study of Ghanaian drumming in the American academy, three of the Council's members were directly responsible for the inclusion and transformation of drumming and dancing into national culture in the 1960s and thus indirectly helped create the conditions out of which Rastahood would later emerge. These central Council members were as follows: J.H. Kwabena Nketia, the renowned ethnomusicologist and founding director of the Institute for Performing Arts at the University of Legon; Philip Gbeho, composer of the Ghana National Anthem and founding director of Ghana National Symphony Orchestra; and George Opoku, founding director of the Ghana Dance Ensemble (the premiere performance group at the Arts Centre).[13]

Under the guidance of the Arts Council, the Arts Centre quickly become a bustling crossroads of performance practice in the independence era. As he told Dor, Gideon Foli Alorwoyie, an Anlo Ewe drummer who, in 1961, Gbeho brought to Accra to assist master drummer Robert Anani Ayitei at the Arts Centre, remembers the diversity of groups that rehearsed there: "all kinds of groups, and they used to call them amateur groups" (in Dor 2014: 240). C. K. Ladzekpo, another Ewe master drummer of national and subsequently international stature, more explicitly told Dor how the Arts Centre's modality of cultural mixing and reinvention transformed drumming and dance into a statal culture in the late 1950s and early 1960s:

> At the Arts Centre, we come to know other groups doing dances from all parts of Ghana. The whole political climate then was an encouragement for us to go back to our indigenous traditions. "Sankofa" [Go back and retrieve it]. It was a very dynamic and a very encouraging time period for us young people. So we picked up the challenge and learned as much as you can from our other neighbors, tribes, [ethnic groups].
>
> (Dor 2014:241)

Among these groups, Foli and Ladzekpo mention the Ghana Dance Ensemble, the Arts Council Ensemble (a.k.a., Arts Center Troupe or National Folkloric Dance Company of Ghana), and the Asiedu Ketete Cultural Troupe (Schauert 2015: 64–65, 149–150, 196–197). Most of these troupes were founded and run by musicians associated first with the Ghana Dance Ensemble, but who subsequently created their own Arts Centre groups as independent side projects, with the end result of transposing the professionalism of the national ensemble to privately run groups. Between them, those who founded these troupes represented the best of Ghana's drumming talent.

Like craftwork, drumming was traditionally learned through tutelage under a master drummer. Gideon Foli Alorwoyie had a classic trajectory. As he explained to Dor (2014:239–240), he exhibited signs of a talent in drumming at an early age. His parents arranged a divination, which confirmed his inclination, and his uncle, a master drummer, undertook his training. By middle school, he was lead drummer. At the 1961 Hogbetsotso festival (Anloga, capital of Anlo), his uncle assigned Foli as master drummer, to display his talent. Husunu Adonu Ladzekpo, the first master drummer of the Ghana Dance Ensemble, heard him play and affirmed the divine prediction. Subsequently, Philip Gbeho heard him and told his father to send him to the Arts Centre when he finished middle school. Arriving in 1964, he began as an assistant drummer to master drummer Robert Anani Ayitei and as a supervisor for some of the Arts Centre's amateur groups. Eventually, Foli would go on to become part of the Ghanaian Dance Ensemble and the Ghana National Symphony Orchestra and subsequently to teach at American academic institutions. Thus, following Polak's model of nationalization as a necessary precursor to globalization, Foli is born into a traditional drumming network, an ethnic, village-based family of drummers. This in turn provides entry into the second, nationally accredited institutional network of the Ghana Dance Ensemble, the Arts Centre, and the Symphony, which yields access to the third, international network of American academic institutions.

This movement from village youth to international ensemble director required several forms of capital: on the one hand, an "authentic," traceable connection to the music-as-culture—attained by being born or brought into lineages of drummers within a specific tradition (Dor 2014:139, see Agawu 2003:97–115, Nketia 1974) and resulting drumming expertise—and on the other, knowledge of and comfort with the English language, Western education, and institutional professional demands.

From Independence and Nationalization to Entrepreneurship and Internationalization

One of the consequences of the move from the local to the international stage was an increasing emphasis on the entrepreneurial side of performance. As the national framework of the 1960s and 1970s shifted to the globalization and privatization of the 1980s and 1990s, the drummers of Accra, like their jembe-playing Sahelian counterparts, molded themselves to the changing socio-economic pressures, shifting over time from being "soldiers of [national] culture" to independent entrepreneurs (Schauert 2015:63; see Polak 2000). They toured and taught internationally and founded private drumming and dancing centers in Ghana, adopting "the national and cosmopolitan practices of the state ensembles as a frame-work for their own business models" (Schauert 2015:135). Hence, even from the early days, it was the institutionalization of Ghanaian drumming at the national level (through the Ghana Dance Ensemble and elsewhere) that provided models for its international transfer, especially to academic settings. Dor ascribes this adaptability to the transferable ensemble model, to Nkrumah's engagement with the diaspora, and to Ghana's political stability and hospitality to foreigners (which has led to American federal grants for international study) (2014:250).

Key to this process was Joseph Hanson Kwabena Nketia, the preeminent ethnomusicologist of Ghanaian (and African) music, who early on, as one of the original members of the Arts Council created by Nkrumah, was key to the establishment of Ghanaian drumming as national culture and as a musical force within the North American academy. As Dor explains, Nketia did this in several ways, beginning in the early 1960s: in his work as co-director of the Ghana Dance Ensemble, in his role as the founding director of Institute of African Studies and the School of Music, Dance, and Drama (later School of Performing Arts) at the University of Ghana at Legon, and in his work at American universities. At Legon, Nketia actualized "Nkrumah's idea of the need to revitalize the deconstructed African culture in the academy" (Dor 2014:245), laying the foundations for an integrated study of Ghanaian performing arts, including theory, culture, and practice. Thousands of students, both Ghanaian and foreign, have been shaped by this study over the decades, and beginning in the mid-1960s, former Legon students taught drumming and dancing in American universities (2015:246). In the United

States, Nketia taught at UCLA and the University of Pittsburg, and he lectured at universities around the world. Both in Ghana and abroad, Nketia's work created opportunities for Ghanaian drummers to work in North America and tour internationally. But of equal significance, it created a flow of ethnomusicologists beginning in the 1960s who came to Ghana to study drumming, returned to the United States to run ensembles of their own, and then from their US base created new flows of students who came to Ghana to learn drumming and dancing, and so on. A cycle of cultural exchange and commerce was thus established (see Dor Chapter 7).

By the early 2000s, this cycle, generated out of the establishment of the School of Performing Art at Legon and of Ghanaian drumming in the North American academy, was bringing visiting college-age Americans (and other aborofo from the global north) to Ghana who were primarily interested in drumming and dancing and related aspects of Ghanaian "culture." These young people were the precise demographic Rastahood would eventually be aimed at. It began with American students being introduced to drumming in college, which then led to their traveling to Ghana for summer drumming programs established by the same drummers who taught them during the year. Subsequently these students might seek a deeper immersion in either Ghanaian drumming or Ghanaian culture, so that a summer trip would be followed by a semester abroad or a stint as a volunteer in the village where they had learned to drum (as was the case with several aborofo I interviewed). Another trajectory led from drumming in college to studying ethnomusicology, conducting fieldwork in Ghana, and entering American academia, thus replicating and expanding the transnational Ghanaian drumming culture in American universities. This too I encountered during fieldwork. Either way, Ghanaian drummers' dominance within American institutions of higher learning from the 1960s onward established a circuit which regularly led young American aborofo interested in drumming and African culture to Ghana, where, by the late 1990s, they were meeting Rastas.

A classic example of this cascade of relations was told to me by Laura, a young oboroni woman I met through her romantic involvement with a Rasta. She had first come to Ghana as part of her ethnomusicology professor's drumming and dancing group, which he brought every summer to study with Godwin Agbeli, one of the elite Ghana Dance Ensemble and All Clans drummers. Agbeli founded Sankofa, one of the popular Arts Centre groups, as well as an arts center, Dagbe Cultural Institute and Art Center, in the 1990s, which supported Kopeiya village outside of Accra through teaching and learning programs for "musicians, professors, teachers from the USA, Canada, Australia, and New Zealand" (www.naniagbeli.com). Agbeli had taught a sizeable group of American ethnomusicologists over the years, many of whom reciprocated by bringing student groups to Dagbe for summer programs, providing volunteers for Agbeli's various village outreach programs, and supporting the center in other ways. Three years before Laura's arrival, her friend had made the trip, where she met one of the youth who worked in Samson's shop. She introduced the boy to her professor, who then started buying all his drums from Samson, and sending all his students to the shop. These foreign students began new exchanges with the Rastas they met at the Arts Centre, exchanges which fanned far out into other forms of labor and opportunities to travel abroad.

Hence traditional Ghanaian dance troupes, established through the Arts Centre in the 1960s to help create a national culture, produced a national and international cohort of master drummers, who found success both in formal institutional settings, like the Arts Centre itself and North American academia, and in privatized entrepreneurial endeavors, touring internationally and then returning home to establish their own private drumming and dancing centers in Ghana. As a consequence, interest in Ghanaian drumming, especially in North America, produced a counterpart aborofo cohort, which expanded the domain of Ghanaian drumming to include American ethnomusicologists trained in Ghana and a flow of students to Ghana pursuing this interest and supporting its drumming and dance centers.

Here the youth of the late 1990s and future Rastas enter the scene. Many of those who grew up at the Arts Centre danced and drummed as children under the tutelage of the local master drummers and internationally successful drumming entrepreneurs, the very people who helped create the flow of visitors that would be so important to future Rastas. These childhood experiences gave Arts Centre youth a significant background in performance. Indeed, once when I asked Moses how long he'd been playing, he laughed and said, "I could drum before I could walk."

"We were born into it," I was told numerous times during conversations about the production and performance of drums, and this was meant literally. Moses was the only Arts Centre resident youth who came directly from a musical family (in which all five siblings have drummed, danced, and taught professionally). Nonetheless, Gideon, Shem, Esau, and Levi had also all trained since childhood as members of adult-run, established troupes that took turns rehearsing on weekday evenings in a building near PWD housing and often gathered afterward at Shem's mother's restaurant.

Moses's list of the troupes that he, his siblings, and his friends took part in as children includes many of the central drumming and dancing groups established at the Arts Centre: West African Folkloric, Folkloric Selamta, Sankofa Dance Theatre, Akroa Dance Ensemble, Adzido Pan African Dance Ensemble, and Ghana Cultural Ballet. The Arts Centre children often started as acrobats, became dancers, and eventually drummers. Gideon, for instance, began as an acrobat at the age of 6 in the troupe West African Folkloric, transitioned into dancing with the troupe at the age of 12, and became a drummer at 17. He was taught to make and repair kpanlogo (Ga), Ewe and Asante drum sets by an older dancer in Folkloric Selamta. Moreover, trained in troupes shaped by a "unity in diversity" national cultural model (Nkrumah in Schauert 2015), the boys understood how ethnic-regional dance forms became desacralized symbols of the nation-state that promoted interethnic integration and subsequently globally circulating, commodified forms of Africanicity. And eventually, they would adapt these tactics, for the transformation and fusion of differing traditions, both in the adoption of their Rasta personas and in the practice of jembe drumming.

But these youth lacked what those who came up in the traditional apprentice system had. They failed in having the proper accreditation (understudying with a master drummer and going through the long years of training). They did not respect ethnic boundaries, as understood by their elders, which produced authenticity through cultural immersion, and this indiscriminacy meant that they lacked the necessary cultural knowledge that would produce a "master drummer." And so Rastas were accused of not teaching "the right thing." Instead their mode of instruction was a mélange—it built on national ensembles but diluted these further, leaning more toward the drum circle model that focused on drumming's role in promoting global unity through inclusivity. This unity, or rather, this dissolution of boundaries between Ghana and aborokyire, was the goal of Rastahood, and thus of Rasta drumming, but it was also a response to the limits of their circumstances, which their elders could not always appreciate.

And yet, both elder master drummers and Rasta drummers recalled master drummer Mustapha Tettey Addy (Mustafa Tadaade in Dor 2014) as someone who embodied a certain shared ideal, although not one that could be accessed by them in the same ways. A figure who bridged the heady nationalism of the 1960s and the entrepreneurial success of the early 2000s, Addy embodied a promise of national integration, and later, of international success. Describing him to Dor, C.K. Ladzekpo recalled Tettey Addy as "not only a Ga. He is a Ghanaian because he knows all the dances" (Dor 2014:241; see Schauert 2015:64). A member of the first generation of the Ghana Dance Ensemble, Tettey Addy founded the Akroah Dance Ensemble at the Arts Centre in the 1960s. In 1969, he joined Adzido, a Ghanaian-American drumming and dancing troupe founded by American choreographer and educator Drid Williams, which established a base in the United Kingdom, bringing Ghanaian dancers and drummers to the country (Schauert 2015:pp). When he returned to Ghana in the 1980s, Addy opened the Academy of African Music & Art (AAMA) in 1988, a drumming and dancing school, performance venue, and hostel in Kokrobite, 30 km west of Accra.

To the young Rastas, who often went to AAMA as both performers and spectators, Tettey Addy was a model of global cultural entrepreneurship, the "happy object," to use Sarah Ahmed's term for "anything that we imagine might lead us to happiness" (Ahmed 2010:29). Such objects become "happiness pointers," creating spatiotemporal horizons that make us "move in a direction toward somewhere *else*, somewhen *after*" (2010:26). In this way, a place like AAMA was an aspirational end goal of Rastahood, which, even if never attained, shaped the orientation of a Rasta life. To have a center like AAMA meant that you had an international career, which enabled travel and adventure as well as a triumphant return home with enough capital to build a house, build a nationally respected and internationally beloved cultural center, and support dependents. Rastas lacked the traditional networks, apprenticeships and elder mentorship that had enabled Tettey Addy to create AAMA, and very few of them ever achieved anything close to it, but it nonetheless provided an aspirational script.

In addition, Addy and the AAMA embodied the outsider image-ideal of Africa, which would become so central to Rasta socio-economic activity. In the center's description, which can be found in various Ghanaian websites, AAMA's creator and marketers fused globally circulating signs of African-icity with Ghanaian drumming and dancing performance and instruction to fulfill the promises of cultural and sonic tourism that Rastahood would also address:

> This *authentic* African *holiday paradise* offers a unique combination of *beach and music*, dance and art. The houses are built in the style of *African* architecture, surrounded by coconut *palms* and *tropical* flowers. In *their own bay* at the beach of the Atlantic, the vacationists and those who are *interested in culture* have the chance to *rest* and *meditate*. The Academy offers classes taught by highly-skilled *teachers* who are all members of the *world-famous* Obonu-Royal-Drum of Ghana group which have *international experience*. Every Saturday and Sunday the Academy is a *meeting point* where you can *experience African music and the local dancing culture first-hand*. A concert hall (up to 200 people) for performances and classes with ballroom are erected in the *African style* of the rotundas and caused by the unique architectural construction which gives optimal ventilation and a wooden dance floor. (my italics)[14]

The italicized words above show how AAMA's promotional materials bring together notions of authenticity, difference, self-care, pleasure, and the culture/nature, body/soul nexus that permeates Global Northern touristic discourse of the Global South. In one circumscribed location, a tourist could find everything they had traveled to Africa to seek: a tropical, beachside holiday paradise (nurturing, sensualized nature) in which to rest and meditate (to care for the self, to be cared for by others); a place in which to encounter and engage with an authentic, local (but "African") culture of architecture, music, and dance (unadulterated, embodied authenticity), but under the guidance of internationally approved and proven teacher-performers (the receiving of knowledge).

Taken altogether, Tettey Addy's trajectory from the independence era through to the early 2000s exemplifies the nationalization and subsequent internationalization described by the many scholars of West African drumming discussed here. The 1990s in Ghana, as we have seen, brought about not only a withdrawal of state funding for culture but also the opening of the country in terms of media, diasporic relations, and tourism. Along with developments in world music, this set the stage for a shift from national cultural production under elder supervision to independent international cultural production and entrepreneurship, under the sign of Africa as imagined by its visitors. Some of those who came up through the old apprenticeship system, like Gideon Foli Alorwoyie, C. K. Ladzekpo, and Godwin Agbeli, but above all Tettey Addy, were able to adapt and transfer their experience to the new socio-economic conditions. These older male models shaped youths' ambitions, but the traditional path toward them, through the patronage and support of elders' networks, was not available. Instead, youth gathered what experience they had in Ghanaian crafts and performance and turned it toward the already internationalized figures of the Rasta and the jembe drummer.

Learning to Play

Just as their training in carvers' workshops enabled them to produce jembes, Gideon and other resident youth at the Arts Centre were poised to play the drum because they had already had musical training as "small-small boys" in the Arts Centre's drumming and dancing troupes. But these childhood experiences, as Gideon explained, had been with Ghanaian drum sets. In its countries of origin, jembe too was traditionally taught via such networks of intergenerational knowledge transfer and accreditation as part of a holistic ethic of living (Melnikoff in Dor 2014:143). But none of the adult musicians at the Arts Centre had been born into jembe networks, as it was not a Ghanaian drum, and regardless, as with woodcarving, the young men lacked the time or support to go through the lengthy process of apprenticeship.

So, just as they had learned how to make the drum together, the boys now learned to play it together. They advanced in tandem by listening to world music albums on cassettes that had been brought from Europe. Despite Ghana's geographic proximity to the Mande region of jembe's origin, jembe music came to Ghana via a Black Atlantic feedback loop monetized and mediated by a Northern/White-controlled music industry.

"We learned by listening to tapes," Moses explained.

Mamadi Keita—he really inspired me. On the tape, I was like, 'What is this?!' It sounds like it's not easy in my ears, but then I thought, 'Okay, let me just give it a try.' It was Levi who was bringing the tapes because he traveled to Austria and Switzerland. Of all [of] us, he was the first one who traveled. So when he would bring it, we all sit together and we listen to the rhythms and we play exactly the same thing. At that time, he was the best one, he played really well, but all of us, all the squad that we had—everybody was good. We were all musicians. So, listening, one person would take one part, another person would take another part, we shared, and at the end, the rhythm was there.

Just as they had modeled their jembe shops after those of the master carvers, using the skills acquired in those shops, the youth now used their training—"we were all musicians"—to learn this new, fascinating drum. They molded a dance-drumming troupe on the AAMA model and other troupes led by the senior men who had taught them as children.

However, again, as with jembe production, youth learned to play the drum to-gether, "away from any disciplinary society": outside of an apprenticeship model that would locate a young drummer within a lineage and immersed him in the jembe's cultural context. In this sense, no matter their skill, they could not become "master drummers," a term that signifies:

> an attained degree of competence, outstanding exhibition of compelling knowledge of a drumming tradition, and a designation of a virtuosic class that the aesthetic evaluators of the carriers of a given culture may confer on a particular drummer. Beyond a demonstrated outstanding level of proficiency in his art of drumming, a master drummer, in some cases, may be regarded as a living repository of his cultural knowledge".
> (Dor 2014: 99, see Nketia 1963; Locke 2009:87–91).

Gideon and his friends replicated the model they had witnessed in their elders. But in taking up drummaking and teaching it to one another, they unintentionally reproduced this form of mobility with a difference. Growing up in the Arts Cen-tre, they had long been witnesses to the organization of shops and ensembles, and to relationships of elder masters and young apprentices. The cultural ethos that linked mastery to adult masculinity remained symbolically powerful, even as its implementation had become increasingly unachievable. So youth endeavored to reproduce this form but sought to evade its limitations, namely by nullifying the long period of dependence and apprenticeship.

Black Star

The same crew of youth who made up Black Star, Gideon's jembe shop, also made up Black Star, the first Rasta-led jembe troupe. The shop/troupe would eventually itself serve as an ideal/model for other youth at the Arts Centre to emulate. Black Star was officially born as a drumming and dancing group when the members of Bamboula 2000, a New Orleans-based world music band came to the Arts Centre to buy jembe drums in 2000. Gideon recalled:

> The first time we performed was in Cape Coast, that was in the year 2000 and we performed with a group from USA called Bamboula 2000 from New Orleans. That was the first performance. And this performance was a jembe session called Black Star ... which we formed from Arts Centre, we the guys making the instruments. We met that group from Arts Centre. So selling the drum to the group, we have to introduce each other: the drum have to talk before the drum can sell. So we talk with the drum [by] playing it together. So after we playing it together, this brother [says], "Oh, I like the drum, the way you play, so let's come together and perform at Cape Coast ... So this the way we come together, through selling drums. So we play at the PANAFEST in Cape Coast, which is Pan-African festival in Ghana.

Bamboula 2000 is a quintessential product of the Black Atlantic; its website describes the band as influenced by "New Orleans, the Caribbean and Africa" (www.bamboula2000.com). Gideon's words situate Black Star's first performance within the history of the proliferation of interest in the jembe across the Atlantic and shows how Rastahood locked into these circulating diasporic resources. When Gideon said, "So we talk with the drum [by] playing it together," he was echoing a statement that has been made again and again about the ongoing conversation between the continent and the diaspora that is music (see Chapter 4).

That first performance's success made it clear that there was an audience for Black Star's youthful, Rasta, jembe version of Ghana. They parlayed that first performance into ongoing gigs throughout the Ghanaian coastal tourist zone, aiming at their niche of the international audience. They found this audience in the coastal strip, the oldest part of Accra (see Chapter 1), which included the Arts Centre, the Riviera Beach Hotel, the popular Labadi Beach to the east, and the performance venues Akuma Village and Osikan to the west. Moving north into the city and away from the beach, a few clubs and bars on the busy Oxford Street in Osu, and the School of Performing Arts at the University of Ghana at Legon also formed part of this map, as were the annual Ga festivals that transformed Jamestown, the Alliance Francaise's monthly music nights, and oboroni parties at homes or hotels, to which youths were invited as either friends, performers, or both. These locations formed a contact zone where young people from Ghana, the diaspora and aborokyire, cohabited around the consumption of music and dance. This contact zone would house the formation I call Rastaworld, recognizable by the particular combination of young aborofo and young, West African Rastas who engaged in the production, performance, and sale of neotraditional African culture.

Moses explicitly linked Black Star's creation to the youths' economic goals and the tourist market as the best place to achieve these: "Black Star was something that we made where we can also survive from it, like to go and play in performances and in the parties. We knew some friends who worked with tourists, so when the tourists arrived in Ghana, they would call Black Star to come and perform in hotels, sometimes at private parties."

Their 2001 performance at Oasis Restaurant in the city of Cape Coast, which I had incredibly witnessed as my first encounter with Rastahood, turned out to have been a pivotal moment for them. At Oasis, Black Star met a group of like-minded Cape Coast youth who were drumming and dancing at a school called Agro. Like the Accra boys before them, the Cape Coast youth brought prior training in Ghanaian drumming to their entranced encounter with the exotic drum.

Moses recalled, "We introduced the jembe to Cape Coast. At first, they were playing jembe a little bit, but they didn't know what it was about until we popped up. Then they saw that, 'Wow, this is really amazing!' So then it became popular-popular-popular in Cape Coast." As the central site for diasporic heritage tourism, the city of Cape Coast provided an excellent audience for drumming and dancing.

Gaining entry into this scene was the happy accident that suddenly expanded Rastahood's reach and its opportunities for mobility far beyond Central Accra.

As with the initial opportunity to play with Bamboula, a drum purchase started it all with the Cape Coast youth. Samson explained:

> A drum brought it all together …—they come to Accra to get a drum, and from that time, we liked each other … I would bring friends, we would play in town. …We would play around the Castle Restaurant the first time [before] it burnt down, then in Agro. So it started getting big; it started bringing a lot more people from Accra. So after some time, yeah, after some time, Kofi [the leader of the Cape Coast crew] too have the mind, it's good that we keep it going, so all the time when we come we stay with them, so then we said, 'OK, let's try something like that in Cape Coast.' … So [the chief] and some people sat down that day we brought a group from Accra, so we said, 'OK, we're starting something for Cape Coast, a Cape Coast version of Black Star.' Kweku had the mind to call it African Drumbeat.

Based in Cape Coast, African Drumbeat eventually went international. By cultivating key relationships, Kofi, its young director, successfully forged connections with several European music schools, which sponsored Drumbeat for European tours of performances and workshops. These initial tours could not have happened without oboroni sponsorship, which included organizing, booking, and promoting the tour, writing letters of invitation for visas, sending advance payments to Ghana for the purchase of visas and airfare, and organizing housing at the destination. Youths often arrived in Europe with little to no money and depended on their partners' support until they could make some.

If owning a place like AAMA or Agro was Rastahood's long-term happy object, traveling abroad on a tour was the short-term goal. There was the sheer pleasure of seeing the world, experiencing difference, making new friends, and feeling the love of new audiences. But also, as part of a troupe, a young drummer could forge new and lasting connections. These could lead to drum contracts, sales of other tourist objects, follow-up solo invitations to perform and teach in Europe, sponsored collaborative productions of CDs, and the sending of new guests and tourists in Ghana the drummer's way. Often, these north-south journeys created crucial friendships and romantic relationships. In this sense, youths' jembe trajectories belonged to a wider category of West African men for whom drumming abroad enabled global mobility and "rooted cosmopolitanism" (Appiah 2006; see Gaudette 2013 and Schauert 2015).

Like AAMA and other centers like it as well as earlier connections between traditional Ghanaian drumming and North American academics, then, African Drumbeat both breached and bridged the boundary between Ghana and aborokyire, creating a bi-directional flow of people, culture, commodities, and capital. A steady flow of Ghanaians began traveling to Europe to perform and teach drumming and dancing, while a reverse stream of Europeans came to Ghana to learn and consume it. A parallel exchange of objects took place: when youths traveled north, they took drums, Ghanaian clothing, and tourist art objects. When they returned south, they brought Nikes, backpacks, cellphones, digital sound recorders, and other commodities.

Rastahood: Vicissitudes of the Happy Object

In September 2004, I traveled to Cape Coast with Samson, Moses and others for Fetu Afahye, the annual Fante harvest festival that marks the turning of year and is the time of homecoming for Fante people (Blankson 1973). It was my first time back since 2001, when I had first met them all in the same city. Afahye drew large crowds of locals, returnees, visitors, and tourists and provided multiple opportunities for performance. It was a time of year when many who had settled in Europe returned for a visit home. Kofi's family house was full of transnational Rastas; Fante, Twi, and Ga were interspersed with Danish and English. Reggae was always playing on the stereo. A long row of Nike and Adidas sneakers lined the hall. Drumbeat or some variation thereof performed at least once a day: on the opening night of the large fair that ran for several days in the city center, and on following days in Cathedral Square, at the family compound, and at Castle Restaurant, the old haunt from our first encounter in 2001.

That summer, Moses and Sampson were moving in opposite directions. Moses had been staying in Cape Coast for a few months, working as the master drummer for Drumbeat. For years, he had struggled to survive in Accra, and he seemed calmer and more confident in this smaller, more affordable city. Through his work with Drumbeat, he had left Ghana for the first time to tour in Europe the previous year. Now, he was preparing for a teaching trip to Denmark. He was feeling sad about leaving, happy at the prospect of earning some money, and nervous about the funds he needed to get onto the plane. As I describe in Chapter 1, Moses had been born into a drumming family and was exceptionally talented. But drumming was all he knew how to do, and performance opportunities in Accra did not earn him much money. He was hoping that this trip to Europe would lead to something permanent.

Samson was ecstatic to be in Ghana; he had recently returned from a difficult time in the United Kingdom and planned to stay for several months. Samson had first entered the Arts Centre as a young hawker and there learned to make jembes with youth who had grown up on site. While he himself had not been raised in the drumming and dancing troupes, Samson was a quick study with a seemingly innate business sense. He had a flexibility and charisma that enabled him to make the most of his opportunities and encounters. The formation and success of African Drumbeat radically expanded the Rasta map, which now, through Cape Coast, was directly linked to Europe. When students came to Cape Coast to learn, they bought jembe drums from Samson in Accra. By 2001, business was good, and a loan from an oboroni girlfriend allowed him to purchase a shop halfway down Carvers' Lane. Here, he developed relationships with foreign dealers (for whom he made large contracts of 100–300 drums). He bought a second shop and brought his family (his cousin and his brother) into the business. By 2003, it was by far the most successful of the Rasta-owned jembe shops at the Arts Centre.

Much as the woodcarver shops had been spatial manifestations of the eastern Volta Region in the capital, Samson's shop made Carvers' Lane part of another map and evinced the encroachment of jembe youth culture and power. While carvers often worked desultorily in quiet shops empty of customers, Samson's shop, centrally located among them, was always buzzing with activity. Music or football from the scratchy radio competed with drums being tested and young men laughing loudly as they worked. When big contracts had to go out, extra "boys" were hired to sand drums, apply drumheads, or pull rope. Piles of neatly stacked drums dried in the sun, and at times, a large container truck parked on the field in front of the shops waiting to be filled with goods to take to the port. When Moses was in town, he spent hours drumming at the shop, which was often the site for impromptu drumming sessions, sometimes accompanied by dancing, which burst out of the shop sonically.

The shop was a central stop-off point for young aborofo traveling in Ghana. They came not only to buy drums to take home with them, but also to hang out with youths and thus to experience Ghana through the Rasta lens. Sometimes Moses taught dread-locked oboroni boys to drum there. Sometimes, an oboroni girl who had gotten involved with a Rasta attached to the shop hung around. Often the shop was the last place aborofo hit before heading to the airport. Much as people from the Ewe villages would use the carvers' shops as a base in Accra, aborofo would drop off their bags, run last-minute errands, and buy last-minute gifts from the Handicraft Market (often with a Rasta facilitating negotiations between customer and trader). When the time came to go to the airport, a quick performance often saw them off.

One of the oboroni girls who arrived at Samson's shop via the Cape Coast connection was Amanda, a student at a British music college who perfectly exemplified the Rastascape described in this chapter. At her college in the United Kingdom, she had taken a Ghanaian dance class, which led her to Cape Coast for a summer dancing program, where she got involved with African Drumbeat as it was getting started. She was brought to Samson's shop to buy a drum, and soon after, they got together. Eventually, they had a child, and Samson moved to the United Kingdom so that the child could be born and spend his first year of life there.

It was Samson's first time out of Ghana, something he, like most Rastas, had wanted for a long time. But as often happened, the realities of life in Europe were much harder than the fantasy. Planning to hold drum workshops, sell drums directly to consumers, and to perform, as people did when they went on tour, Samson had to confront the hard fact that such tours were always executed by youths' European partners, who were often connected to music schools and communities in their home countries. Without such a network, it was hard to drum up gigs or meet potential customers and students.

Reflecting on this that September 2004, Samson explained, "When you talk about the music side, the art side, there are people there too who are musicians and it's not very organized for them to get a job in the music. If you want to work with schools and things, you have to do it through an organization …—but not all the people there are able to have a link to get the work."

It was in these moments of transition from Ghana to Europe or the United States that youths most felt their lack of formal training, institutional affiliation, and membership in professional networks. Living bicontinentally was hard but also essential to Samson's musical and professional development. Each place gave Samson something that he could then carry to the other.

As he explained to me that September in Cape Coast, in Ghana, he "took in" himself with the knowledge, inspiration, and material that he had to "give out" in Europe. He refilled his spirit.

You know, when you are in Ghana, it's when you train and you build up inspiration to get *there* [Europe]. If you are there, you'll be doing it, you'll be doing it … But when you go to Europe—it comes down, the spirit comes down. Because it's different—the environment and everything—it's different, the way of living. … You won't be learning—you will be giving out rather than taking in.

Moses chimed in:

You cannot stay there for the rest of your life, you know. By all means, home sweet home. You come back home again. So if you go there and you work for some time you have to come back home. … You have to come back home and learn more about your culture.

But Europe provided other forms of learning. "Music is what makes my life beautiful here," Samson said. In Ghana, he was never alone. In England, he often was.

I have time to be in the house and study music. That is what I do now, I listen to all kinds of African music. I am learning and thinking, and I know when I go back and do [my own] thing, it will be different. That is the thing about Europe—in Africa, it is very difficult to get musicians from one place to come to another, but when you come here—here you can have the whole of Africa.

It took a few years for Samson to get his grounding, but eventually he did. He was one of only four Africans in his town, a lonely shock at first, but he was fortunate to befriend an older Gambian drummer who lived in a nearby town and who began including him in his workshops and performances. In this small network, he taught, sold, and fixed drums, and traveled to London as often as he could to see and perform with Ghanaian friends and to mitigate his homesickness. He often had to return to Ghana to deal with family or shop issues, which made it difficult for him to commit to long-term programs in the United Kingdom. Meanwhile, his absence from Ghana made managing the shops difficult, and over time, his shop managers took over his contracts, preferring, as Moses had explained, to work for themselves rather than for him, and eventually opening their own shops. Eventually, Samson and Amanda's relationship ended, but he continued to split his time between the United Kingdom and Ghana, successfully parlaying the capital from one place in the other. By 2012, he had parlayed his business and contacts successfully enough to create, with an American partner, his own, smaller version of AAMA in the same village of Kokrobite, close to Accra popular with aborofo and youths alike.

Very few youths were as lucky as Samson, whose talent, charisma, entrepreneurship, and ability to make aborofo care about his projects was unusual, but Samson's story exemplifies the enormous potential of being a Rasta who produced and played the jembe drum. For most others, Rastahood provided smaller returns, which often rested on the ways Rastahood, essentially accompanied by Rasta *style*, generated contacts with oboroni women. This topic will be discussed in the next two chapters.

***

Rastas were the perfect carriers of the universalized form of African drumming that the jembe represented in the West and which so strongly resonated with aborofo. Rastas did not come from a jembe lineage and had not studied under elders from a specific drumming tradition. Like older drummers, they sought connectivity, access, mobility, exchange, but they had to find a new avenue toward these. By traditional standards, they lacked authenticity, musical expertise, mastery of cultural knowledge, and membership in established networks. Instead, they assembled a mélange of borrowed rhythms, not just of Ghanaian traditions but also— critically—of the jembe drum, which had become thoroughly globalized, to create a new persona—the Rasta. Ideally, this figure spawned a cascade of relations of intercultural reciprocity, connection and exchange.

The role of the jembe would not have been unfamiliar to youths, having been schooled in the making of objects that served first rural-traditional, then de-sacralized-national, and finally commodified-global functions. They merely expanded this to include a regional object as part of the integration, the newest resonant symbol of Africanity within oboroni imaginaries. The 1980s and 1990s saw the internationalization and commodification of national culture and heritage under an increasingly globalized, neoliberal capitalism. With the rise of world music and the commercialization of heritage, the jembe drum rose to international prominence as arguably the quintessential cultural symbol of "Africa." In this context, the drum became a potent object that had the potential to generate social, economic, and geographic mobility.

When youths drummed and danced, they produced microcosms of the cosmopolitan, intercultural imagined community they wished to inhabit. The drum called people toward them, when before they had had no way of calling and being heard. The "authenticity" of Rasta jembe music was not musical, but rather intersubjective, lived, embodied, and experiential. As a tourist, you could admire a carving as a materialization of that which had brought you to Ghana, but you couldn't re-carve it or inhabit the experience of its maker. In contrast, the drum, and the drummer, could be repeatedly engaged with, through immersive, participatory spectatorship and learning. Interpolating tourists as participants, a jembe could establish relations that lasted much longer than a craft purchase. It was no longer a simple craft commodity; its sale to an oboroni (foreigner) was not the endpoint of a transaction but the beginning of a relationship. To do so, the drum needed to be played by a particular type of person—a Rasta—who embodied an oboroni fantasy of Africa. This figure is the topic of the next chapter.

Notes

1 In some cases, modern drum shells are made of slats of wood instead of carved from a single log, and the drumheads may be bison or synthetic rather than goat. Similarly, tuning on modern drums can be done through cinch-style or lug or conga-style tuning (Kalani 2002:7–9). The drums youths made followed traditional materials and techniques.

2 See Nketia (1974), Agawu (2003), and Arom (1991)for polyrhythms.

3 Dafora, trained in opera in Europe, founded Shogolo Oloba, the first African dance troupe, in the 1930s and helped found the African Academy of Arts and Research (Charry 2005:9). Dunham, a prima ballerina of the Chicago Opera and a student of Radcliffe-Brown, Sapir and Malinowski, became the first African-American to present indigenous dance on the concert stage and formed Ballet Negres, the first American black ballet. *Africa and the Americas: Culture, Politics, and History*, pp. 535–540. Born in Trinidad and raised in New York, Primus traveled through west and Central Africa learning dance and was a member of Dafora's troupe early in her career (Charry 2005, Cohen 2012, Dor 2014: Chapter 1, Emery 1988, Sunkett 1993, Perpener 2001). See Emery 1988 and Perpener 2001.

4 Polak (2000:18–19) credits specific, spectacular performances by Fode Youla (Berlin 1978), Adam Drame (France 1984), and Soungalo Coulibaly (Switzerland 1986), with the sudden explosion of the drum's popularity in particular locations.

5 The first to come was JHK Nketia (1958), who began teaching at UCLA in 1963. He was followed by Kobla Ladzekpo (Columbia University, 1964; California Institute of the Arts, 1970; UCLA, 1976), Abraham Adzenyah (Wesleyan University, 1969), C. K. Ladzekpo (UC Berkeley, 1973), Alfred Ladzekpo (California Institute of the Arts) (Charry 2005: 13). See Dor 2014 for a history of Ghanaian drumming in American Universities.

6 *Billboard* debuted a world music chart in May 1990 (Taylor 1997:5), reinventing it as a sale tracking category, and throughout the 1990s, a Grammy award category and dedicated magazines and other forms of coverage demonstrate its growing commercial impact (Feld 2000:150).

7 Charry postulates that this transformation must have been profoundly ambivalent for Olatunji, whose goal had been to rebuild lost African traditions in the United States (Charry 2005:15). But Charry also points out that Olatunji's own hybrid music (the fact that he did not come from a drumming lineage and or single tradition) probably made his role in this process possible (2005:17). For obvious reasons, master drummers disapproved of drum circles. Dor cites several ensemble directors who lament the anything-goes ideology, the loss of polyrhythmic structure, and participants' misguided belief that they are performing authentically African music (Dor 2014:123).

8 Ebron (2002) notes that Francis Bebey's *African Music: A People's Art* (1975), a foundational text that, along with Nketia's *Music in Africa* (1974), established African music as a field of study, was key to the introduction of this representation of African music, and thus Africa, in the west, continued by subsequent ethnomusicologists (Chernoff 1979). These essentialisms, which repeatedly racialize the mind/body and thought/feeling split as White/Black or European/African, were also made by African thinkers, most famously Leopold Senghor in his version of *negritude* (cite) (see Soyinka 1988 and Fanon 19xx for critiques).

9 The liner notes to Olatunji's *Drums of Passion*, written by his cousin, sociologist Akinsola Akiwowo, follows this narrative. The drum is represented as a sacred object charged with evocative, mysterious power, life force, and divinity. In line with the authors of negritude, Akiwowo further argues for the universal significance of African traditions (Charry 2005:6).

10 Determining the economics of this trade is very difficult, as prices vary dramatically depending on people's need. I was often told that people would do work, or sell drums, for much less than their value if they badly needed the money. Moreover, even though I was not in the market to buy drums, I was an oboroni, and thus a potential client. Therefore, I can't confirm whether the prices I was quoted are exact. I also was not especially interested in exact numbers. It was clear that people were struggling to make ends meet and earning little profit after all expenses were paid on an object.

   That said, in 2005, these were prices I was quoted for jembe production ($1=9090 GHC [second Cedi]): shell (carved in Okurase, a village outside Accra and transported by bus): 50,000; iron rings (welded at the Arts Centre): 50,000; mostly goat, some cow skins (different sizes, purchased at the Arts Centre by traders from Burkina Faso and northern Ghana): 15,000–20,000; rope (purchased from fishing companies in Accra): 20,000–22,000 per drum; decorating a drum shell: 10,000/drum; sanding (the shell): unknown, probably 5,000/drum at most; mounting (the drumhead): 10,000.

   And in 2012, I was quoted these prices: mounting drumhead and rings ($1=1.79 GHS [third Cedi]): 15 GHS, pulling, sanding and designing the shell surface: 5 GHS, polishing the shell and shaving the skin: 3 GHS. Raw materials, much more expensive than labor, take up most of the payment, leaving little profit and requiring more than 50% of the money up front. A single shell for such a contract would cost 15 GHS. Add transport, the cost of other materials, and labor, and by the time a drum is sold (approximately 50 GHS to a wholesaler, anywhere from 80 to 120 GHS to a consumer), there is little profit for the owner of the contract. Wood typically used was tweneboa (botanical name: *Cordia millenii*; *Cordia platythyrsa*), but also at times nyamedua (*Alstonia boonei*), odum (*Milicia excelsa*), and African mahogany (*Khaya ivorensis*). See https://www.ghanatimber.org/species.php and http://djembe-art.eu/african-drums-wood.htm.

11 I saw this fragmentation and proliferation elsewhere in the Ghanaian artworld at this time. For instance, by 2003, Accra had several art schools (a university degree in art could still only be gained at the university in Kumasi), but the headmasters of these schools had all originally been teachers at GHANATA, the first of the Accra art schools.

12 Due to my entry point into this project via art rather than music, I did not interview elder leaders of traditional drumming and dancing troupes in Accra. My analysis of elder troupe leaders here depends on conversations with people at the Arts Centre and draws heavily on recent ethnography on the topic by George Worlasi Kwasi Dor (2014) and Paul Schauert (2015).

13 As Dor explains, the global preponderance of Ewe-as-Ghanaian drumming is largely due to Philip Gbeho, who regularly arranged Anlo Ewe pentatonic melodies, which the national orchestra performed with "typical Ewe drum-rhythmic

accompaniment" (Dor 2014:238). Gbeho frequently brought young talented drummers from his region to the capital to perform in the symphony, the Ghana Dance Ensemble and the Institute of Performing Arts (2014:239–240). Gbeho further promoted Ewe drumming and dance through the Gbeho Research Group, a drumming and dancing troupe that rehearsed at the Arts Centre itself (where his driver was Moses's father, an Ewe drummer).

14 For instance, this description can be found on Information hood (https://www.information-hood.com/academy-of-african-music-arts-contact-details/) and Ghanayello (https://www.ghanayello.com/company/12020/Academy_Of_African_Music_Arts). Accessed May 23, 2023.

4   Styling the Rasta Self

Consider the following contrast. Among other photographs tacked to the wall of Ezekiel Poku's shop (Ezekiel is a jembe producer introduced earlier), a 2001 photograph, from when he first started drummaking. It shows a chubby teenage boy with a shaved head, wearing a polo shirt, sitting on a bench, smiling hopefully up at the camera, a jembe shell between his knees. In contrast, in a photograph I took of Ezekiel in 2004, he is transformed from good schoolboy to young Rasta. In this subsequent photograph, he sits back, smiling confidently, in a plastic chair, marijuana smoke wafting around his head, his hand gesturing outward, an exemplar of Rasta style in sports sandals, yellow glass beads, *asa saa* pants (patchwork pants made by a local tailor who specialized in Rasta clothing and oboroni beachwear), and a tie-and-dye t-shirt of Ezekiel's own design batiked with a map of Africa. Cowrie shells decorate his jaw-length dreadlocks in a signature look.

Ezekiel's image exemplifies many elements of Rasta style. It pulls objects into a new semantic set, sometimes merely by placing them in combination on the body of a young man in the city. In this case, the style is exemplified by Ezekiel's eclectic aesthetic of abundance, through a layering of materials and references that often include multiple patterns of Ghanaian cloth and the repeated representation of Africa on various accessories, a kind of overload of signification that blasts the viewer.

But Rastahood is not merely a mode of self-presentation. It embodies a multimedia tourist art form that generates mobility, a compound form with different entry points for subjects with distinct talents and skills. The more skills one controls, the greater one's sphere of agency and mobility. It is also a direct replacement for a prior mode of agency and independence. As we have seen, at the Arts Centre and places like it, tourist art and its craft model had, for a time, supported youth in becoming adult men, but the economy of the 1980s and 1990s created economic conditions that precluded most youth from accessing adulthood and respectable masculinity in this manner. They formed part of a generation of African youth struggling in waithood and seeking new ways to access mobility and adulthood. Tourist art nonetheless provided youths of this period with a background in material craft, performance (for those who grew up in the Arts Centre), and tourist relations. In turn, youth applied these acquired skills to the production and

DOI: 10.4324/9780429244568-5

performance of the jembe drum, a new commodity brought to Ghana by the liberalization of the 1990s and the rise of world music.

Hence, jembe production, the object-making aspect of Rastahood, and jembe performance, the formal performative aspect of the art, expanded youth's mobility geographically beyond Ghana and existentially beyond bare survival. But, they were not, in themselves, sufficient to replace the pathway to adulthood and independence that had been lost. Something more was needed to produce a domain of intersubjective agency and a fully possessed sense of self. As we will see, it was style and the work of the body that turned out to be crucial for the production of Rastahood, in the search for a new path forward.

And so, as youth were shifting from elephants to drums, they were simultaneously taking up a new "Rasta" mode of self-presentation, which became available to youth, as the jembe did, during the neoliberal opening and transformation of the 1990s. The wooden masks, elephants, and jembe drums made and sold at the Arts Centre were, of course, signs from elsewhere, filtered through diasporic, colonial, and tourist lenses, but they were continental African signs. Rastafarianism came from farther away and was more fully diasporic in its origins. It was a specifically Jamaican iconography of Africa that had taken on a global life of its own, circulating throughout the world for decades, and was therefore familiar and legible to the tourist gaze: a view of Africa *from a distance*, as a holistic, singular entity. This meant that attaching Rasta style to jembe production and performance gave late 1990s youth a figure that not only potently embodied and materialized "Africa" as seen through the oboroni lens but thereby also produced through oboroni visitors a fertile interface with the world outside of Ghana.

Achieved through everyday performativity and fashion, Rasta style, hence, transformed tourist art by radically expanding the art object to include the artist himself (Harding 2013, Kasfir 2007). Through embodied, stylistic performance, the 1999 generation merged closer and more distant signs of Africa, Blackness, and Black masculinity into a Rasta "mask." In turn, their art practice was one of reading these signs, translating them, envisioning new forms, designing with them, and ultimately making themselves into the very material of these signs, becoming themselves an image of Africa, Blackness, and Black masculinity that provoked and fulfilled oboroni desire (Kasfir 2007:201). To reach young, mobile, empowered, culturally open, and politically liberal foreigners, the style had to simultaneously connote a progressive, open-minded cosmopolitanism *and* a local traditionalism. It had to be strange but familiar, foreign but legible. The Rasta mask thus fused the instantly recognizable aesthetics and practices of the diasporic style of reggae and Rastafarianism (dreadlocked hair, clothing, accessories, language, and gesture) with Ghanaian neotraditional elements (cloth, beads, shells, and leather), to create a locally specific version of a global sign of Africanity and Black masculinity.

It is important, again, to stress the specificity of the Ghanaian Rastas described in this book as those engaged in tourism (as opposed to reggae music or the religion of Rastafari). While there are religious Rastafarian communities in Ghana, as well as a lively local reggae scene, the Rastas in this book were not religious adherents, nor were they (necessarily) reggae musicians. Though they sometimes performed (and always consumed) reggae music, the Rasta component of their practice consisted of fashion, style, and embodiment. Today, elements of Rasta style are pervasive throughout Accra, while Ghanaian reggae dancehall has made its mark both nationally and internationally (Alleyne 2017). But this was not the case in the late 1990s and early 2000s, when hiplife (Ghanaian hip hop) was the dominant musical youth culture (Osumare 2012, Shipley 2013) and Rasta style was marginal.

What is more, the Rasta figure only makes sense interculturally, in the partially overlapping contact zones that bridge Ghana, aborokyire, and the diaspora, where aborofo, Black diasporans, and fellow Ghanaians connect as interlocutors, witnesses, and audiences. Indeed, considered from local, which is to say internal, point of view, Rasta style was, for many Ghanaians, an inauthentic mimicry of something foreign intended to "catch a tourist," to attain mobility through relations with an oboroni.[1] Craftwork, drumming, and dancing attracted both men and women as spectators, collaborators, students, and patrons; it was for the general populous. But the Rastafication of the body, from the perspective of mainstream Ghanaians, was imagined to cultivate desire specifically in the female tourist and specifically for transactional purposes, thereby making Rasta style a disingenuous, sexualized self-presentation that provoked shame in the mainstream Ghanaian beholder (I discuss this dynamic in detail in the next chapter).

But, even if sometimes addressed to the (female) tourist, Rasta style was more expansive, serving multiple functions and addressing multiple audiences simultaneously. It functioned as a call and a welcome to aborofo through its recognizable iconography of Africanicity, and it fulfilled a parallel desire to be welcomed by aborofo. In short, it was a style intended to generate connectivity and relation. And it was in this spirit that Rastas presented themselves, as often the first people an oboroni encountered upon landing in the country. Indeed, Rastas waited at Kotoka airport to welcome aborofo when they got off the plane. They hung out around the cheap hotels and hostels where aborofo stayed. They could be found at the beaches of Labadi and Kokrobite, at the Arts Centre, at the cultural institutes and "spots" where aborofo drank, danced, and listened to live music, and especially, at Akuma Village, the Rasta hostel, restaurant, performance venue, and housing complex that overlooked the sea a mile south of the Arts Centre. In most places where tourists could be found, especially those interested in immersing themselves in the "real" (read non-elite) Ghana, Rastas and their style were present, sensing a certain oboroni desire for a particular experience of "Africa" and through it bringing Rastaworld and its possibilities into being.

In addition, for many Rastas, taking on Rasta style was not only, or even mainly, about connecting with tourists. It was not merely oriented outward, away from Ghana. For many youths of the 1999 generation, Rasta style was about a new generational outlook, joyful self-expression, cosmopolitanism, and rejection of the societal demands of respectability which had failed to secure futurity.

Rastafying the body most explicitly marked the sharp pivot youth made away from the foreclosing modes of Ghanaian masculinity and adulthood they had inherited but could not fulfill. The self-conscious stylization and performance of the body, hence, operated as both a critique of and a solution to the problem of neoliberal (im)mobility.

In the diaspora, Rastafarianism and its iconography had always harkened back to African roots and African power. Now the roots had come home once again, to be taken up in new ways as manifestations of empowerment, independence, and Africanness/Blackness. Rasta style, like Rastafarianism itself, simultaneously celebrated an African past and promised a new African future, in this case to a generation coming of age under neoliberalism, for whom the promises of success under the old life scripts were being challenged by a privatizing economy that forced youth away from collective occupational frameworks into individual entrepreneurial endeavors (Mains 2007, Quayson 2014, Ralph 2008, Shipley 2013, Weiss 2004).

Thus, Rasta style, as it emerged at the turn of the millennium in Ghana, was a subcultural style that looked both ways, internally addressing Ghanaians, externally addressing aborofo, signifying simultaneously in two separate systems of value. Within a Ghanaian discourse, Rasta style uttered refusal rather than respect. It was an oppositional subculture that rejected mainstream norms of comportment, dress, and work, thereby creating a space of social separation for Rastas from the mainstream. Within a global touristic discourse, the style produced an alternative—a solution—to this self-imposed exile, by generating a new sphere of relations with aborofo. Rasta style was well-poised for this polyvalent maneuver because the style's polysemic capacities allowed Rastas to have two conversations at the same time, make two proclamations to different audiences, and engage in two projects of selfhood, via the same stylized body. These two simultaneous conversations operated by drawing on different aspects of a first-globalized and then-indigenized Rastafarianism, deploying through the same signifiers different meanings in Black/African and interracial/intercultural contexts. Insofar as it offered opposition instead of compliance, Rastahood resonated with Ghanaian youth within the local context of their lives. But it *also* increased youths' outward-oriented mobility by enabling relations with aborofo. Thus, to become Rasta was both a sincere and a strategically essentializing choice.[2]

A closer examination of the sources and uses of the Rasta somatic style, then, helps shed light on how it provides a mode of agency, in addition to the craft of the drum, for addressing the problem of stasis and the desire for mobility that the 1999 generation confronted. Toward this end, this chapter describes how reggae and Rastafarianism brought the Rasta style to Ghana and then lays out the components of a specific Ghanaian Rasta style, focusing especially on dreadlocks, the style's most locally salient and symbolically powerful element. The chapter then explores closely how youth deployed the stylized Rasta body to work on two sets of relations, one addressed inward toward mainstream Ghanaian society and the other outward toward an oboroni audience.

The Rasta Style

For Ghanaian youth of the 1999 generation, a visually and sonically striking, globally recognizable Rastafarian repertoire of looks and practices was ready at hand. It included elements such as the Lion of Judah, a vibrant palette of red, yellow, green, and black that colored everything from clothing to accessories, to objects, to architecture; the empowering Rastafari language of Iyaric (Dread Talk); the countercultural practice of smoking cannabis; and above all, dreadlocks, which by the late 1990s, in a global context, transcended their connection to Rastafarianism and served as a free-floating sign of African beauty and pride (and additionally, on the continent, as a sign of cosmopolitanism).

Youth, for their part, localized the Rastafarian image by merging it with a neotraditional repertoire that fused partial aspects of distinct aesthetic practices into a generalized "traditional" African iconicity, which itself had "bush" and antimodern connotations. This neotraditional repertoire included various kinds of Ghanaian cloth (batik, tie-and-dye, waxprint, and poor man's cloth, or waxprint patchwork); beads (especially glass); shells (the cowrie in particular); seed pods; and carved coconut shells—all of which were used to make clothes, jewelry, bags, and other accessories. That the Rastafarian and Ghanaian national palettes were identical, both drawing inspiration from the red, yellow, and green of the Ethiopian flag and from the red, black, and green symbolic of Garveyism and Pan-Africanism, made this integration all the more fluid.

While many of the discrete elements in the Rasta set circulated more broadly in Ghanaian visual and material culture, the Rastahood of the 1999 generation resignified these dissimilar objects into a distinct subcultural style that both produced and was produced by the contact zone. For instance, waxprint cloth is ubiquitous in Ghana, but patchworking uneven strips of it into *asa saa*, or "poor man's cloth," often has countercultural, artsy, Rasta, and oboroni-hippy connotations.[3] Batik and tie-and-dye are also widespread, but the application of the techniques to a t-shirt to produce an icon of the continent (which showed up on a host of Rasta and tourist art objects ranging from jembe drums, to coconut shell passport bags, to jewelry) was distinct.

A classic example of the unfinalizability of objects, the fluidity with which their meaning could be remapped again and again, could be found in the use of the cowrie shell, which defined Ezekiel's signature hairstyle. Arab merchants introduced the white mollusk shell *monetaria moneta* (money cowry), a native of the Indian Ocean, to West Africa in the 14th century. Subsequently, during the slave trade, it became a key currency and later went on to be used as ornaments, charms, and objects of divination throughout the subregion. Eventually, it came to be so localized a symbol that when Kwame Nkrumah introduced a new Ghanaian currency in 1965, in an anticolonial rejection of the British West African pound, it was named the Cedi (Akan for cowrie). Thirty years later, the shell, reentextualized by the tourist trade, was ubiquitous in jewelry worn by Rastas, at which point it signified "tradition" and locality.

For Rastas, then, the surroundings of Accra offered a host of material possibilities for the production of their style. Its various components could be purchased throughout various cities and towns in Southern Ghana, but especially around the Arts Centre, which gathered weavers, seamstresses, tailors, leatherworkers, brass and bead jewelers, etc. It was a circumstance out of which each young man could mix and match, creating his own, unique version of the Rasta style.

The Function of Style

As far as the body is concerned, what is style and what does it do? Writing about clothing and fashion in South Africa, Jean and John Comaroff write, "The body … cannot escape being a vehicle of history, a metaphor and metonym of being-in-time" (Comaroff and Comaroff 1992:79). Similarly, for Terence Turner, the body surface is a "social skin," "the symbolic stage upon which the drama of socialization is enacted" (Turner 2012:12); that is, the surface of the body functions as an interface between the individual and society, the private and the public, the existent and the possible. And for Kobena Mercer, arts that center the body through stylization and performance have been solutions "to a range of 'problems'" (Mercer 1987:35); whereby "distinct, if not unique, patterns of style…are politically intelligible as creative responses to the experience of oppression and dispossession" (Mercer 1987:35). In other words, differing patterns of style not only situate the body in history and the social sphere but also offer a way to recast the order of things.

Part of this power to disrupt comes from the way style operates like a language. Indeed, scholars of African fashion and style have shown how modifying the body through clothing constitutes "a political language, one comparable in eloquence and potency to the spoken words of the most skilled orator or the written words of the most compelling propagandist" (Allman 2004:2). But, in contrast to the spoken and written word, the stylization of the body operates "at a slower pace than speech events" and is thus a more durable agent of change (Hendrickson 1996:15). Medium and message, on the social skin, power is "represented, constituted, articulated, and contested" (Allman 2004:1). New styles thus emerge to disrupt existing power structures, by rearticulating them, be these structures gendered, intergenerational, intercultural, or as is often the case, all of these and more. What is more, styles that generate value, especially the potential to bring something new and disruptive into being, often draw upon foreign sources, be these "Western" or "diasporic," through which to rearticulate the local and familiar, thereby generating their disruptive force (Allman 2004, Gott and Loughran 2010, Renne 1995, Rovine 2015, Rovine and Adams 2002).

Rastas, I suggest then, are fashionable tricksters (Donkor 2016, Shipley 2015), relatives of dandies, flaneurs, and others engaged in performative acts of sartorial agency that combine self-conscious style on a performative body to create "an embodied, animated sign system that deconstructs given and normative categories of identity" (Miller 2009:10). Through their self-conscious stylizations, Rastas interrogate the fixity of normative categories of identity and resignify them by regrouping or reentextualizing objects into powerful new statements (Gates 1989). The

Rasta image-form itself is, thus, composed of such reentextualized sets of object-signs, in the case of Ghana, object-signs which are not only morphologically but also geographically and historically distinct. And through these acts of combination, Rastas thereby composed new clusters of meaning, new political speech acts. Hence, Ghanaian Rasta style exemplifies the "unfinalizability" of objects, their simultaneous durability and mobility (Appadurai 1986, Bakhtin 1990), through this style's appropriation of global Rastafarianism and its merging with a Ghanaian neotraditionalism.

This process of constant remixing—"the often surprising ways in which blackness travels" in and through popular culture (Mercer 2016:227)—has been best theorized by scholars of what Kobena Mercer has termed the "diaspora aesthetic" (Mercer 2016). In a seminal text, Stuart Hall explains, "Always these forms are the product of partial synchronization, of engagement across cultural boundaries, of the confluence of more than one cultural tradition, of the negotiations of dominant and subordinate positions, of the subterranean strategies of recoding and transcoding, of critical signification, of signifying" (Hall 1993:110). What is more, for Hall, these diasporic modes of recoding and transcoding operate precisely within and through the sorts of elements of style crucial to Rastas:

> First, I ask you to note how, within the black repertoire, *style*—which mainstream cultural critics often believe to be the mere husk, the wrapping, the sugar coating on the pill—has become *itself* the subject of what is going on. Second, mark how, displaced from a logocentric world—where the direct mastery of cultural modes meant the mastery of writing, and hence, both of the criticism of writing (logocentric criticism) and the deconstruction of writing—the people of the black diaspora have, in opposition to all of that, found the deep form, the deep structure of their cultural life in music. Third, think of how these cultures have used the body—as if it were, and it often was, the only cultural capital we had. We have worked on ourselves as the canvases of representation.
>
> (Hall 1993:109)

For Hall, then, opening up popular culture, modes of style, and the presentation of the body to serious critical analysis reveals a "profoundly mythic" arena "where we discover and play with the identifications of ourselves, where we are imagined, where we are represented, not only to the audiences out there who do not get the message, but to ourselves for the first time" (Hall 1993:114). It is this position on popular culture and its various elements, one which allows for a view of style as substance, puts a certain primacy on the body, and thereby finds in it an arena for self-transformation, self-discovery, and self-representation to others, which will guide the analysis of Ghanaian Rastahood presented in this chapter. At the same time, in the analysis of Ghanaian Rastahood, it is important to retain a sense of how it produces an identity and not merely a ceaselessly fluid process of reinvention.

Reggae and Rastafari: Global Diffusion and Arrival in Ghana

If we are, then, to understand Ghanaian Rastahood in the tourist contact zones at the turn of the millennium, through modes of cultural pastiche inscribed within processes of self-presentation and reinvention, it is necessary to consider more closely the global and local reservoir of influences from which Ghanaian Rastas drew their practices of embodiment, in particular, the history of Rastafarianism, its styles, values, music, reception, and transformation over time. Although the Rastahood described in this book became a phenomenon in Ghana in the 1990s, reggae music and Rastafarian religion were both present much earlier in various niches of the culture. This made them available for the youth of the 1990s to draw upon— and to combine with other similarly available elements of Rastahood—once the need for a new mode of life arose.

Rastafari, the religious, political, and cultural movement which began in 1930s Jamaica, was a culture of resistance against centuries of oppression, exploitation, enslavement, and colonialism. Among its multiple influences, Rastafari drew upon Garveyism, the Black Nationalist ideology of Marcus Garvey, and Pan-Africanism. Rastafari called for Black pride, the joining of all Black people in brotherhood to fight for the decolonization of Africa, and repatriation—the return of slave descendants to their African homeland. Positing a world in which Blackness was equated with good and Africa (Zion) with heaven, Rastafari offered a racialized political critique of and a solution to oppression within a White supremacist world.

Reggae music in particular, once it exploded onto the international music scene in the mid-1970s, "gave Rasta access to the World Stage" (Bradley 2001:61, King Bays and Foster 2002:Chapter 6, Murrell et al. 1998:Introduction), especially through reggae's most famous musician (and one of the most recognizable people anywhere), Bob Marley.[4] Through the spread of reggae music, Rastafari over the next two decades became "a potent symbol and expression of defiant independence, racial pride and solidarity" (Savishinsky 1994:20) for Black youth the world over and resonated in other communities struggling against centuries of European and American colonial expansion. Indeed, as an ideology, Rastafarianism, "a marginalized liberal movement from a 'little rock'," was apt for transfer to other contexts of disenfranchisement due to its globally resonant goals: its focus on developing "a psychology of Blackness and somebodiness," its rejection of bigotry and classism, its way of "holding onto a messianic hope for the future," and, finally, its creation of "big, big music" that attacked social problems (Murrell et al. 1998:10).

As reggae and Rastafari became symbols of power and hope, businessmen seized on musical the genre as an opportunity. As Neil Savishinsky and Sebastian Clarke point out, record companies entered the reggae music business in part due to the large African market for Caribbean music (Clarke 1980, Savishinsky 1994:22). This global dissemination transformed reggae's message (Edmonds 2003). To make it more palatable to White audiences, Island Records, the record label which had near-hegemonic control over the genre's global circulation, sought to strip Rastafari of its original transgressive celebration of Blackness, as a consciousness ideology (Clarke 1994, Edmonds 2003, Whitney and Hussey 1994). Hence, to distance reggae from its oppositional political roots, Island Records reframed Bob Marley as a spiritual ambassador of racial harmony, mysticism, and universalism, replacing radical Black politics with "natural mystique" and placing Marley in a newly formed music category, "Roots reggae" (Alleyne 1998, Järvenpää 2016, King 1999, Stephens 1998). In this way, much like the jembe drum, Rastafarianism was transformed through reggae into a polyvalent signifier—steeped in oppositional Black politics in certain contexts, generalized as a universally available emblem of idealism and hope in others.

Rastafarianism arrived in Ghana in the 1970s. The first Rastafarian community in Ghana was founded by Jamaican Wolde Mikal in the mid-1970s, and the largest Rastafarian community, the Twelve Tribes of Israel, was set up in the mid-1980s as an offshoot of the eponymous Jamaican organization by its leader, Prophet Gad (Savishinsky 1994:33). By the 2010s, the country housed ten Rastafarian mansions, or communities, all of which were represented by the umbrella organization Ghanaian Rastafari Council (Alhassan 2020, Dovlo 2002, White 2007, 2010). Carmen White (2007) argues that Ghana became a focal point for Rastafarian repatriation because of "Ghana's legacy as a significant site for developments in Pan-Africanism, past and present," especially Kwame Nkrumah's pioneering Pan-African policies of uniting African nations and inviting peoples of the African diaspora to return to Ghana (2007:691). Other scholars have pointed to Ghana's mecca-like site for repatriation for the same reasons (Gaines 2006, Jenkins 1975, Lake 1995, Weisbord 1973).

In the 1990s, President J.J. Rawlings, inspired by Kwame Nkrumah, enacted his own Pan-African policies. This period saw Ghana opening outward officially as well as informally, increasing the presence of both diasporans and diasporic forms in the country. Officially, the exchange between Ghanaians and members of the African diaspora was mediated by the Culture and Tourism Ministry under President Rawlings, who, following in Nkrumah's footsteps, invited diasporans to participate in Ghanaian nationbuilding, this time through the creation of cultural markets within a neoliberal framework (Adrover et al. 2010, Holsey 2008, Shipley 2015, Williams 2015).

As part of this appeal to the African diaspora, in 1992 President Rawlings established PANAFEST, the biannual Pan-African cultural festival, and in 1998 Emancipation Day, which had been celebrated in the Caribbean since 1834 to commemorate the final abolition of chattel slavery in the British colonies. Indeed, Ghana became the first African nation to celebrate the day, as part of a concerted call to people from the African diaspora, even going so far as to call itself the "Gateway to the African Homeland." Ghana claimed this position due to its former status as major coastal exit point for African peoples bound for enslavement in the New World. Many diasporans heeded the call, among them Rita Marley, Bob Marley's widow, who moved to Ghana in the 1990s and settled in Aburi (mirroring in some ways a similar move by W.E.B. Du Bois, who emigrated to Ghana in 1961 at the age of 93, renouncing his American citizenship and living in the country until his death in 1963). This decade of policy culminated with the Ghana Immigration

Act of 2000, which extended special status and rights to people of African descent coming to Ghana from the Americas.

The liberalization of media markets in the 1990s furthered the spread of diasporic music, as nascent private radio and television stations spread reggae and hip hop throughout the country. Through these cultural forms, the African diaspora became a site of cultural innovation *and* continuity, a shared-yet-other space for Ghanaian youth, full of value and potentiality; (re)connecting with these diasporic resources was a way to craft a viable future. By the time reggae reached the ears of disenfranchised Ghanaian youth, in the late 1990s, they welcomed the rupture of Rastafarianism: its message of somebodiness, of universal brotherhood and equality, was a welcome alternative to the classist obstacles to mobility at home and the racist obstacles to it abroad. They also welcomed its appeal to aborofo.

Rocky Dawuni and Ghanaian Reggae

The story of Ghanaian reggae and Afrobeat singer-songwriter-activist Rocky Dawuni ("Ghana's Bob Marley" (www.cnn.com)), whom numerous Rastas named an inspiration, encapsulates the transformative period that began in the 1970s and led to the development of Rastahood by the 1999 generation. His story, as recounted to me in an interview, is included at length for this reason. Born in 1969 into a chiefly family of the Konkomba tribe of northern Ghana, Rocky Dawuni grew up in the military barracks of Michel Camp, where his father was stationed. Michel Camp was the base for the First Battalion of Infantry of the Ghana Army, located in Tema, Ghana's major port, 24 kilometers from Accra. As Dawuni explains, the barracks were "like a cultural ecosystem," with an unusual mixture and intimacy between people of different tribes. Indeed, like the Arts Centre boys, Dawuni learned Ga, Fante, Twi, and Hausa as a child and was familiar with different cultures, foods, and traditions. Codeswitching was thus essential to life in the barracks, "kind of a survival thing, but at the same time it was really interesting, because you just open different parts of your brain." In addition to this multifaceted experience growing up in the barracks, Dawuni's extended family exposed him to Anglicanism, Islam, and traditional religion. Altogether, these experiences allowed Dawuni to have "a deeper, broader perspective of who the African is or was."

Dawuni's childhood in the 1970s was also a time of great political change in Ghana. He was nine years old when Jerry Rawlings attempted his first military coup against the generals who had ruled Ghana for the decade since Nkrumah's deposal. "So I grew up in that era of Ghana, where it was change, it was people power, and also a leader that came and was so loved among the people, because his whole manifesto was to stand for the plight of the underprivileged, and also a defender and fight against corruption."

This exciting period was followed by the government-imposed curfews of the early 1980s, which effectively shut down Accra's vibrant live music scene. But even as the country was closed, music remained accessible in the barracks through

Ghanaian soldiers, connected to the outside world through their participation in UN peacekeeping missions. Most of them were music lovers who always brought back records, "because at that time, you'd come, everybody wants to see your record collection, look at the sleeve and all of that. It was a very prestigious thing." Records also trickled in from family members abroad, increasingly bringing African reggae alongside Jamaican reggae.

High among African reggae artists who garnered exposure in Ghana at this time was Ivoirian reggae musician Alpha Blondy, who regularly performed in the Mande and Baoule languages and who, according to Dawuni, "was also conscious and represented Africa." Kojo Antwi was also prominent, who, along with his bandmates in the Classique Handles (Classic Vibes), appeared in the first Ghanaian band to sport dreadlocks and who, in his solo work, successfully hybridized Roots reggae, Highlife, and Afropop. In the end, though, like many others, Dawuni's first memory of a Ghanaian performing reggae was watching Felix Bell perform, huddling around a neighbor's TV with the other kids from the barracks.[5]

And so, reggae was "around." But the event that caused reggae to fully enter Dawuni's consciousness was the day Bob Marley died (May 11, 1981) and the palpable sense of loss around him. Soon after, Michel Camp's military band was rehearsing and its leader, Julius Mango, entranced him with a song. It was Bob Marley and The Wailers' "Coming in From the Cold" (1980). Dawuni describes an almost "genetic recognition of the sound. The music just spoke to you in a way that you're like, 'Oh this is what it is, but this is how it could be.'" In the three quotes below, Dawuni articulates the three-part argument for reggae's resonance on the continent described above. First, reggae resonated with Africans because it was African:

> First of all, it was understanding the rhythmic and melodic aspects of the music. That was the first thing, is the language. Music is communication. … It's a rhythmic language or a melodic language. When Bob Marley sings, that's like an old African wail that people knew, that he could have picked from somewhere and was his idea of connecting with Africa. … There was rhythmic—and at the same time too the lyrical—connection and invocation of Africa.

Second, reggae was part of the ongoing musical call and response, the "engagement across cultural boundaries" (Hall 1993:110) that continuously generated Black popular culture and the "deep structure of Black cultural life" (1993:109), a recognizable musical language that bonded the motherland and the diaspora together:

> And there was also the whole thing of the reverse communication: speaking back, answering. So we saw our heart as answering that call. So the call was coming, and there had to be an answer, and the answer had to be musical. And to answer had to be acknowledging that we heard what is being spoken, what is being talked about. … it was also seen as a process of reconnection: the diaspora to the motherland, and the motherland to the diaspora.

Finally, there was reggae's universality, how, like the jembe, reggae was simultaneously about Black experience and a profoundly resonant form for people everywhere, defined a universally applicable truth and intention:

> Most of the artists who were singing, they didn't see themselves as singers. They saw themselves as spiritual vessels who were speaking a certain truth, who were representing a certain human condition, and trying to make the world better, first of all for their people, and then ultimately for everybody. … There was nothing about reggae to serve like, 'Oh, let's go make money, let's make this.' … Its intention was spirit: connect to God, connect to nature, be good to others, uplift others, live for others. …Those were the noble ideals that I felt, because those things too were true across people and races and color and gender and all of that. It became true to everybody. And it allowed the music to resonate with everybody who heard it, because it communicated to them on a level that was higher than just the music that they were listening to. I think that was reggae's success.

Hence, for Dawuni, reggae offered both a shared sense of community, through a universal Blackness and Africanness, diaspora and motherland together, each inextricably intertwined with the other and connected by a basic musicality that transcended reggae itself; and, at the same time, reggae offered a sense of a higher truth, one which looked outward toward others, toward God and nature, and thereby uplifted his own people and others in a broader shared "human condition."

And yet, alongside these diasporic and universal connections, Dawuni recognized how reggae also had a more specific and local significance. He saw that the transAfrican and transdiasporic elements of reggae, along with the appeal of its higher truth, provided young Ghanaians with their own generational way to reinvent their identities.

> So as a young man growing up in the 80s, that was what we were hearing and seeing. And it was also a celebration of the strength of Africa, because we were dealing with that also: what is the future, who are we identity-wise, after eras of Ghana being first African country to be independent, … from our first president to military rule by generals, and then the coup era, and then the 1980s? So the 80s was, 'Okay, we're going to remake ourselves again. Who are we? What defined who we were?'

That was the 1980s and Dawuni's youth, the discovery of reggae and the call toward a new sense of Africa, the diaspora, and a binding, uplifting, higher truth. By the early 1990s, Dawuni was studying philosophy at the University of Ghana at Legon, where he put together his first band, Local Crisis with fellow university students. At the same time, he met his future wife, business partner, and producer Cary Sullivan, an American who was doing the university's study abroad program. Dawuni's university career coincided with Ghana's transition from military governance to multi-party politics and deregulation, which ushered in the privatization and subsequent proliferation of radio, television, and other media. Reggae could now be heard on the radio, with show hosts, including Black Santino, Culture B, Daddy Bosco Sefa Kayi, and Black Moses playing the music to Ghanaian audiences and making up almost 20% of radio airplay (Alleyne 2017:102, Chude-Sokei 1994, Savishinsky 1994:24).

Following graduation, Dawuni and Sullivan led a bicontinental life, working in the music business in Los Angeles and Accra. But Dawuni purposely returned to Ghana to record his first album, *The Movement* (1995). With this choice, Dawuni made himself part of a generation of cosmopolitan Ghanaians returning to the changing country. Recorded with Local Crisis, *The Movement* was produced at the formerly government-run Ghana Film Studios, which would eventually become Ghana's free-to-air network TV3 in 1997.

According to Sullivan, "It was a very exciting time, very cutting edge, we were all just making it up as we were going along because there was no blueprint for what we were doing, in any way." Indeed, producing music in Ghana at the time was a novel and precarious enterprise, because Ghana lacked the musical infrastructure of its Francophone neighbors. "Every African artist that you heard internationally [at that time] came from a Francophone country because they were set up to go through France," Dawuni explains. "You got signed to a French management company, they push you to Europe, a French label—you got all of that. We did not have that, so it was a lonely path, where you had to kind of anticipate and create your opportunities."

The two were able to transfer knowledge they gained in the LA music industry to Ghana's growing media industries, eventually producing albums, music videos, and live concerts (including the highly successful Independent Splash Festival, which they ran at Labadi Beach as well as the National Theatre and other venues). A reggaescape was growing in coastal Ghana, as found in Rastafarian enclaves at Labadi Beach, Akuma Village, the seaside village of Kokrobite, and the town of Aburi in the Akuapem Hills north of Accra. Sullivan, herself, became head of entertainment at Metro TV, Ghana's first independent television station, and while there produced *Smash TV*, a nighttime magazine show, on which Ghanaians could, often for the first time, watch music videos from elsewhere on the continent.

Notably, *Smash TV* also played the polished videos of a young Reggie Rockstone, who was, alongside Dawuni, the musical celebrity most admired and most often named by youth in the 1990s. Hence, even as Dawuni was creating his fusion of reggae and Afrobeat, which he would eventually call AfroRoots, Rockstone, recently returned from the United Kingdom, was introducing young Ghanaians to hiplife, a term he coined for the new, Ghanaian hip hop, in which he rapped in local language.[6] Born in London to Ghanaian parents, Rockstone was educated in Ghana but returned to London in the mid-1980s, where he became involved in the nascent hip-hop scene. Returning to Accra in 1994, Rockstone, like Dawuni, was a "culture broker" (Barber 1987, Hannerz 1992, Shipley 2013), an agent of indigenization (Osumare 2012), who brought everything he learned abroad back home and translated it into something refreshingly local and thrillingly cosmopolitan at the same time. Looking at Accra "through the eyes of a global traveler with local knowledge" (Shipley 2013:81), Rockstone and Dawuni embodied the at once diasporic and homecoming Ghanaian who was so crucial to the cultural renewal of the time.

Magical Hair: Dreadlocks and a New Identity for Young Men

Hence, just as Tettey Addy and Olatunji personified the ideal life made possible by drumming, Dawuni and Rockstone embodied a type of cosmopolitanism that offered a new identity and possible avenue toward success for youth in the 1990s, and they did so by creating localized forms of music and style created in the diaspora and popularized with youth everywhere. Dawuni and Rockstone were young, educated citizens of the world, who lived, variously, in Ghana, the United States, and the United Kingdom, and who stylistically and musically represented elements of these faraway places through their creation of "local, blended genre[s]" (Shipley 2013:40). In so doing, they bridged the distance between aborokyire and Ghana. And, importantly, they did so while bearing a visible form that encapsulated, for themselves and others, all of their experiences: the undeniable, immediately recognizable, and potent mode of outward self-presentation found in long, dreadlocked hair. Through them this singular sign of Rastafarianism would undergo a crucial process of resignification in Ghana.

Hair was a particularly powerful element of the "social skin," a frontier of the social self, upon which changes in status or the violation and inversion of norms could be enacted, thereby made visible, and consequently read by others (Turner 2012:486). In his classic essay "Magical Hair," Edmund Leach (1958) argues that hair is a cultural symbol—a public culture—in which a shared language of actor and audience makes the goal of communication possible. Moreover, hair is an in-between substance, midway between nature and culture, the individual and the collective, "erupting from the body into the social space beyond it" (Turner 2012:488). As Kobena Mercer states,

> [H]air is never a straightforward biological fact, because it is almost always groomed, prepared, cut, concealed and generally worked upon by human hands. Such practices socialize hair, making it the medium of significant statements about self and society and the codes of value that bind them, or do not. In this way hair is merely a raw material, constantly processed by cultural practices which thus invest it with 'meanings' and 'value.'
>
> (1987:34)

Rasta hair enveloped all of these functions. It was a readily apparent symbol, the first thing one might notice about a person, that bore its social significance outwardly and demanded to be read. On the other hand, the quality of hair as signifier allowed dreadlocks, like the rest of Rasta style, to communicate various messages simultaneously. And yet, at the same time, the in-betweenness of hair, bodily manifestation and sign, naturalized its cultural work, enabling Rastas to embody (rather than, for example, speak) what they represented in particularly powerful ways. Indeed, dreadlocks in particular, which were often described by Rastas as "natural"—what African hair "naturally" does—cleaved to claims of essence so strongly that it was as if they were in themselves their own justification, even though, like other elements of the Rasta repertoire, the social life of dreadlocks both before and upon arrival in Rastaworld was long and varied.

Locks were thus simultaneously very old and local and very foreign and new. Hair was so central that the term "Rasta" often meant "dreadlocks" (rather than a more encompassing reference to the style, the ideology, or the religion as a whole). More than anything else, dreadlocked hair visually marked Rastahood and one's commitment to the style. Names, clothing, speech patterns, and behaviors could be (and were) shed and taken up depending on context. Dreadlocks had the slow temporality of commitment. They could be hidden (under hats or cloth) but still made their presence known (through the kinds of hats or head-wrapping styles they necessitated). Therefore, the decision to dread one's hair was a decisive point, preceded by a period in which other forms of Rasta knowledge and practice were acquired. It indexed a fundamental shift of orientation, away from a Ghanaian model of respectability ("correct"ness), in which locked hair served as the symbolic repository for many of the larger negative associations of Rastahood, and toward the more affirmative and empowering resignifications of dreadlocks in the diaspora.

Rastafarians began to cultivate dreadlocks in the 1940s, due to their African origin and biblical reference, and dreadlocks became entrenched in the movement by the 1970s. Jamaicans were initially inspired by images in the Jamaican press of Africans (Gallas, Somalis, Maasai, or Mau Mau, depending on the account) wearing the hairstyle. But as they embraced dreadlocks, they also invoked, as a justification, Levitical law, which prohibits Nazirites from trimming their hair or shaving, as a way of consecrating themselves to the Hebrew God.

In Jamaica, the earliest dreadlocks are attributed to either the "guardsmen" of Howell's Pinnacle commune as an expression of their fearsomeness or to the radical young Rastas of the House of Youth Black Faith as a direct, antisocial assault on Jamaican society through a defiance of social norms concerning grooming (Edmonds 1998:31). This defiance of social norms was an extremely controversial move, which split the movement in two, due to the association of unkempt hair with "mad dialects" and "outcasts" (Chevannes 1994:158). Nonetheless dreadlocks became such a strong visual marker of Rastafari that early police raids on Rastafarians involved head-shaving. Eventually, dreadlocks became "the most salient and visible symbol of Rastafarian identity" (Edmonds 2003:31) and "the most readily identifiable signifier of a meaningful difference" (Hebdige 1979:34).

As in other religious practices, grown, uncut hair, in Rastafarianism, physically and externally embodied spiritual strength and Godly force and was believed to link Rastas to their God (Edmonds 1998:32). Additionally, dreadlocks marked a rejection of Eurocentric standards of beauty, which were central to racist ideologies (Mercer 1987:39), and thus reconstituted pride in one's African physical characteristics, marrying "Blackness to positive attributes" (Cashmore 1979:158, Chevannes 1994, 1998). "Locks spoke of pride and empowerment through their association with the radical discourse of Rastafari which, like Black Power in the United States, inaugurated a redirection of black consciousness in the Caribbean" (Mercer 1987:40). Like other elements of Rastafari, dreadlocks "invoked a concept of 'nature' to inscribe Africa as the symbol of personal and political opposition to the hegemony of the West over 'the rest.' [They] championed an aesthetic of nature that opposed itself to any artifice as a sign of corrupting Eurocentric influence" (1987:40).

Locks in Pre- and Post-Reggae Ghana

Before the advent of reggae and Rastafarianism in Ghana, dreadlocks had two connotations: on the one hand, they were associated with traditional priests and healers, a link between the hairstyle and African spirituality that Rastafarianism affirmed. On the other hand, dreadlocks were the "unkempt hair" of mentally ill, unhoused people who could not care for themselves as proper persons, a con- nection that had also been made in the Jamaican context. Hence, prior to their association with Rastafarianism, long, matted locks signaled "the state of being unwashed, frightening, mysterious, and forbidden" (Ross 2000:167). They were, in the Ghanaian context, a trait of fetish priests and priestesses, royal execution- ers, and madmen, dangerous subjects engaging in roles or behaviors that defied common rules of comportment (McLeod 1981:64). Indeed, among the Asante, the matted hair of priests was called *mpesempese*, "a term sometimes translated as 'I don't like it'" (McLeod 1981:64). Dreadlocks, thus, defied Ghanaian norms of beauty, grooming, and social/aesthetic containment; they marked outsideness and deviance.

Even after reggae and other Black music resignified dreadlocks in Ghana, the association with outsideness remained. Dreadlocks instantly labeled a Rasta as a certain *kind* of person—potentially mad, violent, dirty, homeless, criminal, or ad- dicted to drugs—for whom people had rather low expectations. Dawuni recalls the expression of such biases, when, as a child in the early 1980s, he saw images of Kojo Antwi, a beloved Ghanaian Highlife, reggae, and Afropop superstar, and his band, the Classique Handles (Classic Vibes) in newspapers. They all had locked hair, and for many Ghanaian kids, this was the first mainstream representation of the hairstyle. "And everyone was like, 'mpesempese,'" he said, translating the disapproving statement as "unruly hair." Dreadlocks inspired complex of nega- tive associations, Dawuni explains. They signified a dangerous foreign element, "a rebel person." "Like Bob Marley, they smoke marijuana," which "made people go crazy." Dreadlocks were associated with "somebody who is a homeless person on the street," with "people who are riff-raffs and people who are uneducated and don't have any aim in life," "with no vision and no education." "Oh, you see," peo- ple would warn, upon seeing someone with dreadlocks, as though these heralded a concatenation of terrible events.

Similarly, the carver Ametewe once told me, "You know, Rastas are a new thing here. It has not been long that they are popular. In my childhood, when we saw a man with hair like that, we were afraid." "They think you are mad," Samson confirmed one afternoon, as he wrapped his dreadlocks in white cloth, in preparation for joining his mother at her church. It would be disrespectful to show up in church with dreadlocks, he explained. It would cause "talk" and trouble for his mother. I could imagine—I often witnessed mainstream disapproval of Rastas' hair in unsolicited comments from strangers, usually from older women in public spaces who bemoaned why such a nice-looking young man would ruin himself this way.

And yet, while the above view was still operating in Ghana in the 1990s (and well into the 2000s), the presence of Dawuni and Rockstone added a new meaning to dreadlocks, especially for youth. As Dawuni explains, "[W]hen some of us started gaining national prominence, …it changed the dynamics in a deep way. That argument was defeated, diffused, by that visual and people were seeing it. Now, the young people also saw that there was a certain aspect of success that was attached to that image."

Dawuni and Rockstone modeled a new script; they fused diasporic and local cultural knowledge into a new form; their hair embodied a new "strategic integration" (Askew 2002) of local and global cultural resources that produced a new imagining of what it could mean to be Ghanaian. Dawuni, in contrast to stereotypes about people with dreadlocks, was "university-educated, articulate, and knowledgeable." He embodied "global connection, global success" and had gained national recognition. Moreover, Dawuni accomplished all of this in a way that was especially important for young Rastas facing a changing economy, and he did it by "working outside of the paradigm." Dawuni continues:

> Being bold enough to jump outside of what everybody was accepting and at the same time, finding success in challenging the archetypes and stereotypes. I felt that my success, in addition to other people who wore dreadlocks at that time, became something that inspired the youth to also boldly stake their claim on what was influencing them. To boldly show that identity and be comfortable.

Hence a new set of associations attached themselves to dreadlocks over the course of the 1990s: global and national success achieved by cosmopolitanism and working outside the paradigm. At the same time, it was a rooted cosmopolitanism (Appiah 2006), one which not only brought in influences from outside but also inflected them deeply within the local culture. "I had traveled to America, lived in America, came back, was talking about ideas and things that I've seen and bringing it back and also talking about a sense of morality that was inspired by things that I've been engaged with here, and a global perception and perspective that I've been influenced by." Both Rockstone and Dawuni returned to make music that was defined by a fusion between local and diasporic forms. "I never saw myself leaving Ghana," Dawuni explains. "I saw myself as traveling to acquire knowledge…as you move, there's new stuff that you acquire that re-fines you and you grow. … and then bringing all of that back again, to be part of Ghana's progress and development just like Kwame Nkrumah did."[7] Hence, like Pascal Gaudette's (2013) jembe heroes, Dawuni was Ghana's reggae hero and Rockstone its hiplife hero, widely traveled but rooted cosmopolitans, who brought all the knowledge gained on their travels home and translated it into both national and international economic, social, and personal success. This was the Rasta dream, and Rastahood was more flexible and adaptable for youth who lacked traditional opportunities.

Why Rasta Style?

In 2017, I asked Noah, "Why Rasta style? Why did you all take it up back then?" His answer evokes the layered, local specificity of meanings attached to the figure of the Rasta in Ghana. At first, like Ametewe, Noah assumed that by "Rasta" I meant dreadlocks, so his initial response addressed the hairstyle and its uninter-rupted continuity as naturally African and traditionally Ghanaian.

> Rasta has always been an African thing. Back in the old days, I remem-ber when I was in the North, we had Rasta people, but they were natural Rasta people. They were born like that. They didn't comb their hair; they were priests. When you are born into that family—that clan has a god that they worship, so their god or their spirit prevents them from combing their hair. It wasn't a fashion thing—you were a fetish priest, because of the spirit in you. So when you have dreadlocks, they ask you, "Are you a fetish priest?"

For Noah, as it was for Ametewe, the "Rasta people" of the "old days" contrasted with the new way of being a Rasta, it marked a difference between and coexistence of the spiritual and "the fashion thing," between the African and the diasporic. In Noah's next origin story for dreadlocks, Rastafarianism appears as part of Ghana's opening up to tourists and diasporans in the 1990s.

> In the 90s, we also had a lot of visitors coming from the Caribbean and America for Emancipation Day. Louis Farrakhan used to come with African Americans and they were so proud to see us. They said, "We are coming back home." Rita Marley had actually relocated to Ghana and built her house in Aburi. All this started happening in the 90s. In a way it just sort of changed Ghana, the youth, it changed all of us.

But, for Noah, the most important story is the one recounted by Dawuni above, in which cosmopolitan Ghanaian musicians return to Ghana in the 1990s and provide a new model for young masculinity. Speaking specifically of Reggie Rockstone, the revolutionary Ghanaian hiplife musician, Noah provides a youthful fan's perspec-tive on Dawuni's narrative. Dreadlocks, the most notable aspect of Rockstone's appearance, were its key signifier, and yet, Rockstone was not and had never been a Rastafarian. Instead, his locked hair exemplified how, in Ghana, the hairstyle represented a way to be connected to the world. Noah explains:

In the '90s, Reggie Rockstone came along, and he was using dreadlocks as style. During the old days, it was all highlife. In our generation, in the 90s, highlife was boring, it was fading away—it was outmoded, we called it *colo* [colonial, old-fashioned]. American music took its place and over-came everything. It became so big in Ghana that if you turn on the radio, it's either R&B or hip hop. Reggie had lived abroad and seen things, and now he was back and he was doing his Ghanaian thing, but different. He transformed hip hop into hiplife: he started rapping using [Twi], the Ghana-ian language. He came with this style of rap, and he had dreadlocks, and he was a fashion designer as well—he did so many things that influenced our generation.

Hence, Rastahood was part of a larger, generational cultural formation: like hip-life and AfroRoots, Rastahood sought to express both cosmopolitanism and lo-cality, global belonging, and patriotic pride, and to provide alternative means to achieve professional and economic success under a politico-economic regime which was rapidly foreclosing previous avenues to independence and adulthood, such as had been found in craftwork and traditional drumming performance. Noah explains:

Reggie was trying to give youth the confidence that you can do things for yourself. You don't have to go to university or have some degrees to do something for yourself. A lot of people saw Reggie doing his thing with his dreadlocks it was so nice. It helped so many youth—if it wasn't for Reggie bringing that style, a lot of guys would have been out of work.

Dreadlocks were, thus, intimately intermingled with both existential and pragmatic improvements—with confidence and with the new opportunities that emerged from this confident (self)production. For young men coming up in the late 1990s and early 2000s, without higher education, hiplife and Rastahood were the best of a set of limited options. The two subcultural styles shared a dual purpose: they offered an oppositional, critical stance, for the new generation to mark its own difference and make its own way, and they provided a viable economic model for success and adult masculinity, through cultural performance. Reflecting on the op-positional stance that Rockstone and Dawuni helped make possible through hiplife and Rastahood, Noah remarked:

The way they were growing their hair—it looked so beautiful! We should embrace that we are African; we should stop adopting the foreign style. So these two guys really dominated something in Ghana. All of a sud-den, all this American music just started disappearing—they started just playing Ghanaian music. …We all just started dreading our hair, so dread just picked up!

Suddenly, youth had a way to create a positive space for themselves, in a socio-cultural context that seemed to be leaving them behind. Ultimately, for Noah, and others, Rasta style in Ghana and dreadlocks in particular were about an opening up through looking inward—an opening of Ghanaian society, of their generation, and of themselves. It was about Sankofa ("Go back and get it")—the Akan adinkra symbol so beloved by Rastas, of the bird who turns its head to take an egg off its back. What this bird meant to the Rastas is expressed in the philosophical proverb usually attached to the symbol, *Se wo were fi na wosankofa a yenkyi* ("It is not wrong to go back for that which you have forgotten"), that is, the good of the past must be brought into the present to build a positive future. For Rastas, what was significant was not simply adopting styles that came from elsewhere (the dreadlocks of reggae, hip hop in the form of hiplife, or Rastafarianism), in order to find for themselves out of foreign influences a new mode of self-production, what was significant was the profound inflection of these styles and values through and within existing Ghanaian culture, in order to recast traditional modes of life and values in a way that opened, rather than foreclosed, possibilities for life, in a rapidly changing socio-economic climate at the turn of the century.

Rasta Style in Action: Refusal and Ghanaian Respectability Politics

In a photograph taken in 2004 at the head of Carvers' Lane, Jeremiah, one of the jembe producers mentioned in earlier chapters, wears a shirt, belt, wristband, and lanyard in Rasta colors, as well as a Ghanaian cowrie necklace, and carries a padded waxprint jembe bag. In separate 2004 photograph, taken in his drum shop at the Arts Centre, Jeremiah leans neotraditional, wearing a batik "up-and-down" outfit and cowrie shells, though his woven tam, lanyard, and sunglasses signal cosmopolitanism. In a later, January 2005, photograph taken at Akwaaba Restaurant, the Rasta-dominated restaurant at the head of Carvers' Lane, he wears a camouflage button-down shirt over a crisp white tank top, perfectly ripped blue jeans, aviator shades, Rasta wristbands, well-coiffed dreadlocks, and a cowrie shell necklace. Together, each of these seemingly mix and match outfits, through their heterogeneous elements, materializes Rasta cosmopolitanism, combining cosmopolitanism and tradition, diaspora and Africa, the timely and the timeless, into an image of desirable Black cool.

A 2015 photograph perhaps best encapsulates Rasta style's resignifying fusion and Jeremiah's particular aptitude for this mode of styling. At the time, Jeremiah commissioned a seamstress to make him a classic Ghanaian up and down (shirt and pants/skirt of matching cloth) made of a pervasive tropical cotton print, but he had the top cut like a Northern men's smock, or *batakari*. This commission combined three non-Rasta male uses of cloth (the up and down, the batakari, and a familiar print) into something unique and distinctly Rasta. Leather sandals, glass beads, dark sunglasses, a straw fedora with an Akan adinkra symbol, and a waist-length necklace complete the outfit. The necklace, made of beads carved out of wood, from which a wooden ankh is suspended, is perhaps the most potent sign of creativity and bricolage, with its integration of beads, the Egyptian ankh (a common diasporic sign of Afrocentrism) and, most originally, the use of woodcarving techniques, materials, and scale, which turn the necklace into a wearable work of art. Jeremiah's ensembles traversed the territory of Rasta style, moving along the spectrum from the neotraditional to the cosmopolitan, effortlessly pulling from the various veins in the Rasta repertoire to create varied but coherent looks.

Jeremiah had adopted Rasta style as a teenager precisely to flout conventions. Already, as a student in his hometown of Bolgatanga in the Upper East Region, he personalized his "tea and bread" (the brown and tan school uniform) with a belt made out of plastic water sachets. He commissioned seamstresses to combine rice sackcloth and kente cloth into suit jackets of his own design. He cut and resewed his shoes. This redefinition of mundane objects into "a form of stigmata, tokens of self-imposed exile" (Hebdige 1979:2) meant Jeremiah was frequently getting into trouble with teachers, Scout Troop masters, and other authority figures, who understood his fashion for what it was: style as Refusal, dandyism as Resistance, the upending of convention, and above all a declaration of independent will. And so, from early in his life, Jeremiah adopted a set of sartorial habits that made him stand out and gave him a commensurate reputation.

"I always loved to dress different, just different way, what my soul tells me, what my brain tells me to dress, just to feel what I feel," Jeremiah explained. "And then through that, the name [grew]. 'Who's this guy, he always dresses different? O, that's Jeremiah.'"

Josiah, another Rasta, made the same connection between subcultural style and freedom from convention. Indeed, he literally cut up and remixed the signs of conformity, combining waxprint with jeans or matching female with male clothing. As he explained, these choices were driven by a spirit of contrasts and singular difference:

I see Rasta to be—what [other] people don't like is what they like. That's how Rastas go. They always want to turn out with the unique thing that no one else will put on. And I think that is more attractive—that uniqueness. I for one—I had my jeans shorts, and I cut my sister's dress into pieces and joined that cloth against my jeans to turn my shorts into trousers.

Noah, for his part, made the link to freedom explicit and saw Rastas as doing the work of cultural liberation, which subsequently extended into the mainstream:

[The style] helped us to grow up into our inner being. In Ghana, you can go to university, study all the books, but still know nothing, because of the culture—the culture just teaches you how to do everything. You don't come out of yourself; you are just a follower. You might go to school, but your eye is not open, you are limited. So, we were calling it freedom. Now Ghana is changed, so in a way, we brought democracy to Ghana.

Rasta style spoke a fundamental refusal of Ghanaian values through a defiance of Ghanaian aesthetics, ideas of beauty and propriety, and respect of self and other. It was deliberately cultivated and deployed, serving an oppositional political function, as it had done in Jamaica when it first emerged.

And indeed, Ghanaian youth had something to oppose—the unfulfilled promise of the received script, which said, if you play your role and follow the rules of the system as respectful, obedient youth, you will attain the accouterments and status of adulthood. This conformist position had a stylistic correlate described as "correct" and "neat"—a look which included closely cut hair; clean, starched, ironed, unfaded, normative masculine clothing; a dust- and sweat-free appearance; and contained affect. It communicated respect: for oneself, for others, for the rules of society. However, as we have seen, by the late 1990s, the promise of the "correct" and "neat" way applied to fewer and fewer people. This left many youth fending for themselves.

And so, as a subcultural object, Rasta style stood in direct opposition to the "correct" look. Rastas' fashion, behavior, and affect were "loud" and defiant of local conventions, their challenges were clearly heard. They reflected the will of young individuals, unmoored from a system of social support, striking out on their own. Dreadlocks in particular encapsulated the opposing claims of rebellious youth and those against which they rebelled. To grow and wear dreadlocks was an affront, a deeply oppositional statement, and, crucially, an individualistic act, a sharp turn outward away from Ghanaian aesthetic and ethic norms, into the intercultural sphere, where the individual used the body as a canvas to make their re-orientation known to all parties and subsequently as a starting point from which to seek out new social relations. It is thus with great understatement that, when I asked Jeremiah why he had grown his dreadlocks, he responded, with what was for him typically mellow indirection: "Because I wanted to." This answer elided the onslaught of opposition he had faced by pitting his will and desire against the obedience of social norms, as well as the shifting socio-economic conditions that created pressures to find a new way.

Oboroni Desire and the Rasta Image

Like Ghanaian reggae and hiplife, Rastahood combined diasporic and local genres to produce a new, viable identity, rhetoric, art, and entrepreneurial employment, for a generation confronting privatization and globalization. But while the former addressed African, diasporic, and global audiences, Rastahood's audience was primarily oboroni; it was an objectification of African and diasporic forms for foreigners not of African descent, hence the usefulness of a tourism framework.[8] Rastahood was therefore viewed with deep suspicion by mainstream Ghanaians who expressed their disapproval through a discourse of authenticity. When these

Ghanaians looked at Rastas, they saw a bricolage of decontextualized elements recombined to create a falsified, simplified, homogenized image of Ghana, its purpose to dissimulate, manipulate, and "catch a tourist." To such Ghanaians, their suspicions were confirmed by the confounding, but clear, reality that certain aborofo, in particular young, White women, were drawn to Rastas and sought them out over other kinds of Ghanaians. Thus the Rasta style offended in two ways: by embodying the shameful trickery of Rastas and by reflecting the touristic, primitivist image of the country.

What this disapproving view ignored was the socio-economic predicament in which many Ghanaian youth of the late 1990s found themselves. As noted, Rasta style was a call to be noticed by both Ghanaians and aborofo, a quest to stand out against the norm, for that norm and its promise had failed Rasta youth and they were crafting an alternative path. If Rastahood's first stance was one of symbolic refusal in the face of the failed promise of their elders, for stable adulthood in the form of traditional masculinity, its second move was a pivot outward to solve the problem of this broken promise, a reorientation toward a new interlocutor with a distinct ethic/ aesthetic. Consequently, for this new addressee, Rastahood as outsideness was as desirable as it was repellent to most Ghanaians, "the 'natural' Rastafarian was simply the fetish in the flesh" (Stephens 1998). This was, after all, the reason certain aborofo came to Ghana in the first place, a quest for an idealized and fetishized alternative to the everyday life they were (temporarily) leaving behind (Urry 1990).

The aborofo who sought out Rastas varied—they included backpacking tourists (Ghana on a dollar a day), volunteers, non-profit workers, students abroad, and people who made a living in African "culture" (buying and selling Ghanaian crafts, arts, and musical instruments). But they were united by a politics that rejected elements of EuroAmerican modernity, and late capitalism, in which the sign of Africa served a particular role. They held a perspective which merged an anti-elitist celebration of and desire for the "common people," an appreciation of "traditional" practices, and a utopian, pacifist cosmopolitanism of universal brotherhood. The two components of the Rasta image—Ghanaian neotraditionalism and global Rastafarianism—presented this utopian "Africa" in ways that, for aborofo looking in from without, paradoxically enabled their fusion into one potent emblem, much like the jembe in the drum circle did.

Aborofo looking for an alternative to or temporary escape from their everyday lives found a powerful set of new possible experiences associated with Rastahood, which blended countercultural defiance, traditional culture, mystical force, and naturalness into an overall antithesis to EuroAmerican civilization. Rastahood accomplished this by "turn[ing] the text upside down" (Hall, in Grossberg et al. 1996:103) through rearticulation. Everything that was celebrated in the West was dismissed; everything that had been lost was refound. Hence, Rastafari's original promise to disenfranchised Black Jamaicans of a return to Africa now had a strangely resonant, new appeal to highly mobile, privileged oboroni subjects. In turn, for the Rastas themselves, having composed the Rasta figure, they could now propel this figure out into the world, into new contact zones, where it could act and do and work, for them, in a variety of ways.

Rasta Practice: Mobility and Strategic Masking

With the Rasta figure in hand, late 1990s youth, as Rastas took "control of the image" (Willis and Williams 2002:xi) they projected and to which they were subjected, transforming it into a new, intercultural mask, or persona, designed and made, like a work of art, with specific communicative goals and particular interlocutors in mind. In this manner, Rastas simultaneously became embodiments of an essentializing oboroni desire and agents of change in their own lives (Kasfir 2007:12). Sidney Littlefied Kasfir's argument for the extension of the concept of masquerade to contemporary Samburu warrior practices of self-transformation can be applied to the Rastas own newly devised practices of self-presentation. What is crucial in this conception of masquerade is the way the body itself becomes the artifact, leaving accessories, coiffures, and body language to operate as elements of a "mask," an artistic vehicle or practice of disguise and transformation (Kasfir 2007:238).

Thus, as masquerade, Rastahood garnered its polyvalent quality, offering different masks for different audiences. It enabled Rastas to look both ways, engaging multiple interlocutors at the same time through the complex speech act of style. In this way, taking up Rastahood was always strategic, which meant that it worked at times, but not at others, in certain contexts, but not others. To take control of the image, to don the mask, to self-consciously engender the gaze of the other, these are metaphors for thinking about Rastahood as partial, ephemeral, strategic, yet sincere performances that make sense in particular moments in time (see Harding 2013, Jackson 2005).

Hence, for aborofo, the Rasta figure replicated a pre-existent image, embodying it in flesh. However, unlike an image that was purely a product of aborofo projection, this figure did not stand silent and still, availing itself to the gaze of the beholder. Instead, Rastas "assigned it a place and a role in the sphere of speech and action" (Azoulay 2012:23). The image, hence, both uttered the exclusion of Rastas from the public and, because it was the Rastas themselves who uttered it, overturned the exclusion and produced a new public sphere, according to the Rastas own needs and desires. In other words, the figure of the Rasta operated both inside and outside of dominant culture, allowing youth, through this figure, to both participate in and transform the otherwise limiting context of their lives. To be sure, the knowing nature of this taking control of the image does not cancel out the cultural and racial politics of its figuration within oboroni culture, its instantiation of racialized, colonial tropes. Nor does this knowing appropriation deny the emancipatory aspect of becoming Rasta, a style through which youth express freedom, pride, individualism, creativity, and joy. All of these modes can and do exist together in the performance and production of self through subcultural style.

To explore this model of strategic masking, or masquerade, it is useful to consider two stories of Rastas and the way they moved fluidly through the world. The first is 14 years long and disentangles the meaning of hair, that powerful marker of Rastahood. The second is 14 hours long and focuses on the more easily changeable elements of style: clothes and accessories. Despite their differing temporalities, both stories concern pivots that mark new life chapters and that bring Rastas into contact with new audiences. Together these stories help reveal the ways Rastahood signifies differently from person to person and moment to moment.

As we have seen, Rastahood was a product of the contact zone—it enabled mobility, first into, then through, and finally out of that zone, but it didn't necessarily work outside of the contact zone, in Ghana or aborokyire. Nonetheless, the choices to "put on" or "take off" Rastahood were not inauthentic, or insincere—they were informed readings and strategic responses to the changing contexts that make up any life, but especially one in the contact zone. Hence by putting aside objectifying discussions of authenticity and dissimulation and instead taking a long view of Rastahood as a moving target, ethnography helps bring into relief precisely the quality of sincerity in question here, of a heightened awareness of what is appropriate to one's present circumstance, of a way, through masquerade, to discover intercultural, intersubjective, strategic solutions to the continually shifting conditions of one's life. From this point of view, Rastahood functioned not only as a youth culture but also as a tool to transition out of youth into adulthood by facilitating contact with aborokyire and its potentialities for transformation; at the same time, the mobility thus achieved forced youth to learn to respond to different things in different contexts. What might help you in one life moment didn't help you in another. Once one reached the other side—of the life chapter or the border—Rastahood took on new meaning and had to be put aside or transformed again.

Fourteen Years: Samson's Hair

As one example, let us return to the singular potency of hair in Rasta visual discourse and of how changing dreadlocks marked pivots into new life chapters. We will see that growing, shaving, or re-dreading were all markers of transition and reorientation.

When I first met Samson in 2001, his dreadlocks reminded me of the biblical Samson's hair, a visceral embodiment of his power, a key site of his sexualized masculinity, and a major repository of his attractiveness to aborofo. A vocabulary of gesture surrounded his dreadlocks, which he flipped, worked, gazed through, tied back, decorated, and clothed with great artfulness. It was difficult to imagine him without them.

In 2006, I visited Samson in England for the first time. He was living in a shared house in a small town close to an art school. As one of only four Africans, he was well-known—he sold drums, taught workshops, and performed with a drumming and dancing troupe from a nearby town. On the last afternoon of my visit, we were walking up the High Street with his friend Ebai, a Cameroonian who had lived in England for several years. In gray slacks, a blue fleece top, beige leather shoes, and with close-shaven hair, Ebai stood in contrast to Samson, who wore an earthtone batik tunic with gold embroidery around the collar, a waxprint button-down shirt, baggy jeans with embroidered pockets, and a thrifted black battledress coat with epaulets and brass buttons. His locks almost reached his shoulders. A yellow bead bracelet was on his wrist.

"I almost shaved my head," Samson said.

"It would make your mother very happy," I said.

He said that sometimes he didn't do it precisely because of how happy people would be if he did. The previous week, in London, a Ghanaian woman at a bus-stop had talked to him at length, uninvited, about how he should cut his hair.

Ebai said, "You should do it."

Ebai used to have dreadlocks too and said British people pigeonholed you because of them. Putting on Rastahood got you to Europe, but once there things changed. The contact zone where dreadlocks worked was a narrow space. Outside it, in Europe as in Ghana, locks produced assumptions. Having crossed over to aborokyire, Samson now saw the other side of everything: the good of Ghana, the bad of England, the gains and losses of coming and going. If dreadlocks had brought him to the United Kingdom by presenting him as a certain type of person, he now saw the other side of how that person was perceived. In the end, Samson didn't cut his hair, at least not in 2006.

Three years later, both of us back in Accra, we met at Asasepa, a Rasta-owned, vegetarian restaurant near Akuma Village.

"You look *abafresh* (just come [back])!" he teased me, referring to my paleness in the absence of the Ghanaian sun.

"So do you," I teased back. Samson had gained weight and wore more "correct" khaki slacks, a checkered, short-sleeved button-down shirt, and a Panama hat. But beneath the hat, new, short dreadlocks were growing back.

"I have changed," he said, smilingly reading my eyes. He had; from youth to adult man.

Samson was back in Ghana after three years in England to set up an art program for children, a long-time dream made possible now by years of hard work and a partnership with an American non-profit. It was for the US trip—the embassy interview, specifically—that he had shaved his head. US embassy visa interviews were always stressful unknowns, even if you had all your paperwork in order, and people often shaved their heads to make a better impression, for in this context, dreadlocks were a potential obstacle. Samson's hair made me think of Abel, a Rasta we both knew from the Arts Centre, rubbing his naked head in frustration while playing checkers, describing a failed US interview for which he had similarly cut his hair, which had taken years to grow.

Ultimately, by 2015, the last time I was in touch with Samson, his intercontinental professional life was on solid ground, a sphere of production, instruction, sale, and performance. As a reflection of this successful entrenchment in the cultural field, his dreadlocks were shoulder-length.

Fourteen Hours: Moses' Clothes

When Moses first introduced me to Rastaworld in 2001, he was 19 and already a brilliant drummer. He was fully immersed in Rastaworld but had not yet become a Rasta: he neither smoked, nor drank, nor wore the style, nor talked the talk.

We saw each other again in 2003, both of us coming back from aborokyire, I from New York, he from his first prolonged stay in Europe. While there he had been touring with a drumming and dancing group for several months. When I saw him, instead of the tall, skinny boy I remembered, who loved crisp white shirts and laughingly refused to grow his hair, I was now in the presence of someone much bigger, wearing flashy new Nikes, and donning a head of short, neat locks. A year later, in 2004, he was headed back to Europe for another long tour of performances and workshops. His last day in Ghana (following our week at the Afahye festival described previously) is memorialized in my photo archive, a visual document of the spaces, transitions, and contextual performativity of Rasta style.

Moses had many errands to run and people to see, before boarding the plane that night, and had dressed carefully for his last day. I met Moses at the Arts Centre to find him dressed according to the latest Accra fashion in baggy jeans, a black t-shirt, a crisp yellow button-down, sneakers, and trendy shades nestled in his hair. While there, we went to a seamstress' shop to collect clothes he had commissioned, which he would sell in Denmark: black cotton up and downs, printed with rows of small red, gold, and green Africas, alternating with rows of Black dancing figures and palm trees on red, gold, and green. Seeing the cloth Moses had chosen, for the clothes he commissioned, it struck me how it encapsulated so many of the elements of Rastahood: from the moment of its making, in which a designer had fused Rastafarian colors, the icon of the continent, and neotraditional African signs into an objectification of an existing image, to the moment of Moses seeing and selecting it at the Arts Centre as a pidgin object, to the moment where it would arrive on the body of an oboroni, for whom wearing the cloth would enunciate a counter-cultural, cosmopolitan European identity. Moreover, the contrast of Moses in his Accra clothes, standing next to the elements of the Rasta image he had composed and would soon take on himself, clearly highlighted the way that Rastahood worked as context-specific, strategic masking.[9]

From the seamstress' shop, we slowly made our way out of the Arts Centre and through Accra, stopping to collect some jembes, and then to say goodbyes at a restaurant, an outdoor foosball game, a neighbor's yard, and a relative's house. A photograph taken as we made our way through Accra shows Moses leaning against the bright yellow roof of my car, smiling broadly, in front of a pink building on the mantel of which is painted, "SOMEBODY's ACHIEVEMENT MAKES SOME PEOPLE ANGRY," a reference to the jealousy that good fortune, including contact with aborokyire, can elicit. Moses has often been the victim of such jealousy from other young men in similar straits who were not Rastas, or who lacked his talent as a drummer and the connections to aborokyire that drumming and Rastahood afforded him. It was a coincidence that I parked in front of the sign, but the opportunity it created to capture the image of Moses' hopeful smile, with that message of envy behind it, seemed to encapsulate another aspect of Rastahood.

A subsequent photograph was taken at Akuma Village, a base camp for many Arts Centre Rastas, during the evening. Here we were fully in Rastaworld—gone were the typical Ghanaian clothes and hairstyles of the day. While Samson and others wrote letters to European friends for Moses to deliver, and others milled around in the garden outside waiting to say goodbye, Moses packed. And he had changed. He now wore one of the newly commissioned "Africa" shirts and had replaced the aviator shades with a Rasta tam of red, gold, green, and black. In preparation for his arrival in Europe, he had put on his Rastahood and transformed himself into a sign of Africa.

Rastahood, Mobility, and Style

As a form found in the contact zone, which is a physical space (where Ghanaians and aborofo come together), a symbolic space (where distinct cultural repertoires collide and are differently taken up by participants), and a transitional space (between life in Ghana and life in aborokyire), Rastahood is inherently mobile, enabling travel. Along the way, perhaps more than any other element, style manifests the contextual, intersubjective nature of the art form of Rastahood, by creating a Rasta self capable of continually shifting and adapting—an in-between self along the lines of what Michelle Ann Stephens calls the "inescapably intercultural" *intercorporeity* of the performer, born of his interaction with his spectating other (Stephens 2014). This is why Rasta style was also always a strategy deployed in specific contexts. In this way, the image of the Rasta was inherently mobile and produced an ongoing mobility in the world. It was a way out, a way in, and a way back.

Moreover, as we have seen, through Rasta style, youth were able to address both Ghanaian and oboroni audiences simultaneously. The same performative act conveyed two messages, as part of two exchanges, enacting a critique and offering a solution. Hence, disenfranchised Ghanaian youth welcomed the rupture of Rastafarianism and the lines of communication it opened. Its message of somebodiness, of universal brotherhood and equality, was a welcome alternative to the classist obstacles to mobility at home and the racist obstacles to it abroad. Indeed, the duality built into Rastafarianism-via-reggae made it especially appealing. Its politics within Black communities *and* its mystical appeal to White audiences made Rasta style especially apt for engaging the intercultural position and struggle of Ghanaian youth, by enabling Ghanaian Rastahood to articulate contradictory aspirations at the same time. And so in Ghana, as in other tourist borderzones, Rastafarianism worked on three levels: as a resonant representation of self for disenfranchised Black youth of the urban underclass, as a recognizable representation of powerful Africanicity for tourists, and as an answer to the existential quests of both Ghanaian and oboroni youth. The latter two in particular were commodifiable and explained the presence of Rastafarian youth cultures in tourist zones throughout the Global South.

Notes

1 See Shipley (2013) and Alleyne (2017) for similar mainstream Ghanaian views on youths' appropriation of hip hop and dancehall as inauthentic mimicry. Critically, in these two cases, the uptake of Black diasporic musical forms was not attributed to a desire to attract aborofo.
2 Barbara Olsen's (1995) longitudinal study of the changing role of Rastafarianism over 20 years in the lives of six Jamaican subjects also tracks changing relationships.

3 Although *asa saa* cloth has Rasta connotations, the patchwork cloth is an old Akan textile method that also shows up in high and diaspora fashion (Molly Keogh, personal communication). Schoss (1996) describes a similar patchwork cloth being used by "beach boys" and tourists in Maundi, Kenya in the early 1990s.

4 See Clarke (1980) and Davis (1982) for reggae's emergence as a Jamaican pop music genre in the 1960s.

5 Mentioned by Dawuni, include Sonny Okosun, Majek Fashek, Victor Eshiet, and The Mandators, all of Nigeria; Lucky Dube of South Africa; and the lesser known Malian Askia Modibo. If these were important reggae musicians at the time, according to Dawuni, he also names Evi Edna Ogholi (Nigeria), Amakye Dede (Ghana), and Kojo Antwi (Ghana), as musicians who integrated reggae with other musical styles and thus, in their own way, spread the rhythm across the continent. Equally of the moment, according to Dawuni, was Fela Kuti's Afrobeat, "a music that had a vision just like reggae had." Another key generator of the Ghanaian reggae scene, in Dawuni's account, was Roots Anabo, a Ghanaian reggae band formed in Berlin, which returned to Ghana in 1986 and mentored the next generation of musicians. See Alleyne (2017: 93–95) for a similar history.

6 For the literature on African hip hop, see Saucier (2011), Osumare (2012), Charry et al. (2012), and Shipley (2013).

7 Dawuni was one of the first Ghanaian musical artists to promote social initiatives through his music, a trend very much in place by the early 2000s. "Music, activism, visuals and positivity, in a way that it was more like change through osmosis," he explains. Dawuni is a UN Goodwill Ambassador for the Environment for Africa, a UN Foundation Ambassador for the Clean Cooking Alliance and an Ambassador of Born Free USA, and the Global Ambassador for the World Day of African and Afrodescendant Culture globally recognized by UNESCO (https://rockydawuni.com/).

8 The overwhelming majority of tourists whom I witnessed engaging with Rastaworld were not of African descent. Therefore, this ethnography does not include diasporic Black perspectives on Rastahood, which merit further study. However, there is a literature on Black perceptions of white obsessions with commodified and/or sexualized representations of Rastafarianism which speak to this issue that shapes my thinking in this chapter (e.g. de Albuquerque 1998a, 1998b, King 1999, and Stephens 1998).

9 Writing in a Kenyan context, Sydney Littlefield Kasfir describes a parallel situation, in which young Samburu men traveling to coastal Mombasa to engage with tourists in the early 1990s wore blue jeans, t-shirts, and sneakers for their trip to Mombasa, so as to blend in with other Kenyan men and to avoid being ridiculed as primitive. Once in Mombasa, they changed into their traditional wear, for traditional fashion was part of an assemblage that attracted tourists. This strategic, stylistic codeswitching enabled Samburu youth to use their bodies as "theater" (Kasfir 2004), as a living work of art. Working with Samburu men some 15 years later, George Meiu (2009) provides a related account, in which he discusses the romantic and sexual components of the touristic assemblage, which I describe in Chapter 5.

5   The Affective Labor
of Crafting Freedom

An Intercultural Romance

Brooke first arrived in Ghana as a foreign exchange student from the United States in 1998. At Legon University, she studied at the School of Performing Arts, to which many foreign students interested in Ghanaian culture were drawn. A trained musician, self-described hippy, and agnostic, she came to Africa in search of alternative social, spiritual, and artistic modalities. During her trip, a semester abroad, she encountered two Africas, one confounding, but another which quickly matched the image of Africanicity she carried with her and expected to find, Rastaworld.

> I was really confused at first. I was sticking out, not because I'm a White person, but because I'm a free spirit. I have dreadlocks and I want to smoke pot and I want to play my drum. I was overwhelmed when people wanted to invite me to church and asked me what type of Christian I was. That felt like an offense at first: the vigor with which they requested that I go to their church and the little bit of judgment it sometimes went with when I didn't want to go. Over time it became an irritating piece of the culture for me, it was a lot of pressure.

I understand Brooke's use of the term "Christianity" as a gloss for both religious and other elements that define mainstream Ghanaian society, which were not what Brooke sought or expected to find when she came to Ghana. For her, "Christianity" was not open, not free; it sought, in Brooke's imagination, to constrain her and reign her in. But, like many short-term tourists, she quickly came upon that other, Rasta Ghana. It spoke to what she already understood herself to be.

> I was really into drumming when I was back [in the US]. I was totally bohemian, and I had dreadlocks and the whole thing. And I was really into the mythology, the myths and stories of West Africa. I had taken a class in theater where I had learned about the creation stories on the continent, and I had created a performance based on the Yoruba gods from Nigeria, the symbolism, the colors, the foods, the rhythms associated with those gods. So, I was

DOI: 10.4324/9780429244568-6

really into the mysticism. So, I was going for that, I was so hungry for that. I wanted to know the mystic parts. I wanted to hear about shrines [and] priests who could tell the future. I was looking to *smell* it and be in it and *feel* it and be in a land where that kind of magic worked, and people believed in it. And I just got it straight away by meeting Rastas. Because they were the ones that were connected with that.

Semantically joining drumming, dreadlocks, West African mythology, and mysticism with her countercultural identity at home, Brooke traveled to Ghana to have a direct, embodied, authentic, all-encompassing experience of alterity and found it. And so Brooke offered up precisely the sort of language—one of desire, in the form of a hunger to touch, smell, and hear the magic, to be immersed, transported, and transformed—that embodies this book's theorization of the role of Rastahood and the figure of Africanicity in oboroni experience.

In Rastas, Brooke found a salutary alternative, because "they were the ones that were connected with" the mysticism and the magic. Brooke was introduced to Rastahood when she wandered into the Shrine, a cluster of benches situated in an open field between the university and the main road, blocked off from public view by the grove of trees surrounding it. Like other parts of Rastaworld (especially Accra's Akuma Village), it was a protected space where the standard rules of Ghanaian society did not apply. Brooke explained, "That was where all the Rastas and anybody who was sort of into that open-mindedness would hang out. And…if you wanted to smoke, that's where you would go. A lot of music was made there. A lot of people would bring drums and xylophone and guitar. …And people just left it alone."

Hence the Shrine combined African music, marijuana, Rasta style, and an intercultural milieu into the "open-mindedness" Brooke was seeking, and in Rastas and their polysemous style, Brooke found her Ghanaian counterparts and the Africa was looking for. Here was the "happy object" (Ahmed 2010), which stood for a cluster of promises (Berlant 2011), materializing a cosmopolitan open-mindedness and political critique, at the same time combining this mindset with traditional knowledge and practices, and in so doing forming an alternative to ordinary life.

The first Rasta Brooke met was an artist who embodied the alternative lifestyle she valued back home and the mindset she sought out in Ghana:

He reminded me of my hippy friends at home: really young, jubilant, just a happy hippy with dreadlocks. But he used language like 'Rastafari', 'Jah bless', and all the colloquialisms. For me, Rasta was so much less about Rastafarianism [than it] was a way [of living] for those people who did not buy into the brainwashing, 'cause I felt like the men that I met, not all of them were religious Rastafarians at all; in fact, [only] the minority were. …. This was their way of latching onto something that rebelled against that. I think it's a version of what traditional African priests and religion embodied, but it was just a more modern twist on it maybe. So, there was relief in hanging out with Rastas. And also, we could kind of make fun of that together. There was that solidarity. That's why it was so attractive to me. I didn't come there looking for Christianity.

Here again, in Brooke's language, one finds the combination of a complex mesh of values and styles, understood to be expressed through Ghanaian Rastahood and linked to an idea of Africa. Rather than perceiving Rastahood as tied to Rastafari, Brooke, like the Rastas she met, viewed it, on the one hand, as a locally specific authenticity, a contemporary continuation of the precolonial African religious and cultural practices, and, on the other hand, as a global style, which Brooke understood to be shared by her and Ghanaian Rastas, a marker of modern cosmopolitanism and a countercultural politics that opposed oppression, organized religion, and conformist brainwashing. In addition, for Brooke, Rastahood's party lifestyle, playing and dancing to music and smoking cannabis, manifested the jubilant freedom she wished to experience in Ghana.

After her initial encounters with Rastas, Brooke met Elijah, a talented traditional dancer, who, she felt, especially embodied the values of authentic Rastahood, conjoining, in the particular way she imagined, traditional and contemporary lived experience, along with a certain idea of Africa. Brooke explained:

He had that jubilance and that joy and the music, but there was an anger too. He had a hard life. He grew up very poor in Nima, in the ghetto of the ghettos of Accra. Elijah actually comes from a deep line of traditionalists. His maternal grandfather was a traditional priest and had dreadlocks, so I think he always connected with that.

She was instantly attracted to him. "It wasn't even the Rasta thing that I noticed: it was his pride, his African pride, that most attracted me. It was that he believed he was a beautiful African man, and he wore African clothes and embodied what it meant to be a true African." In his fashion, carriage, and his professional and life choices, Elijah embodied the "Africa" Brooke had traveled so far to find. Again: "it wasn't even the Rasta thing that I noticed; it was his pride, his African pride."

Additionally, there was Elijah's contained affect and the confidence it expressed. "I was definitely attracted to him mostly because he was not overtly wanting me. He was unlike a lot of the other Ghanaian men that I had met, who were obsessed with wanting to be with me and every other White woman." His lack of interest seemed to affirm an authenticity in him, as though his lack of interest in oboroni women (as a category, rather than as individuals) correlated with a rejection of and disinterest in aborokyire. She experienced his emotional and sexual containment as tied to his African pride, anticolonial politics, and self-sufficiency. His affect was part of his style.

Brooke soon moved off campus to live with Elijah and a Rasta friend of his, in a small complex with no running water. Elijah became a portal into the entire life she had desired and imagined as part of her experience in Africa. "It was 100% local life. I just felt like I was living a life that nobody [else] like me was getting to experience. My first two years I had no oboroni friends. I was in a bubble. I was immersed so deeply."

The bubble Brooke Imagines here is telling. It is the opposite of Boorstin's tourism as pseudo-event (1964), a simulacrum geographically located within, but radically removed from, the everyday realities of the place visited, which offers an experience of leisure and wealth shielded from the world around it. In its place was what she assumed was "100% local life, defined by an authenticity which rested upon the notion of access to the back regions" (Goffman 1974), whose "mere existence, and the possibility of their violation, functions to sustain the commonsense polarity of social life into what is taken to be intimate and 'real' and what is thought to be 'show'" (MacCannell 1976:591). It was a different sort of bubble, an immersive experience of alterity that separated her not from Ghana but from others like herself, from her own everyday reality. This particular point of view was not unique to Brooke. Mary, another oboroni woman I interviewed at length, described a similar experience as "my anthropological experiment."

Living with Elijah also offered Brooke and opportunity to survive on limited resources and thereby expand the domain of authenticity available for her to explore.

> That first two years I didn't really work. I was just living off the savings I had. And we traveled. We had a car and a truck and we drove all over. We took a trip all the way to the north to Bolgatanga and bought goats and a monkey. Traveling was inspiring for him too. He was fascinated by the diversity of languages and cultures that existed. When we would stop and eat local foods and explore, it was just as exciting for him. We had so many stories.

In this way, being with Elijah made Brooke feel free and gave her unique access to the "100% local life," to the backstages where authenticity resided. More broadly, in generating "so many stories," Elijah enabled Brooke to collect the kinds of experience and knowledge that were key forms of value for her.

Being with Brooke in turn made Elijah economically free to experience his own country as an adventure. They were perfect partners in adventure, sharing a similarly unfettered interest in exploration. She paid, and in exchange, he provided the immaterial labor of managing their traveling life. It felt like a fair exchange, with each of them putting in what they had. He planned their itineraries, sourced and negotiated their accommodations, food, and other needs, translated the language and culture for her, and generally mediated all her encounters with Ghana. Their bond grew. After Elijah nursed Brooke with great dedication and kindness through a week-long bout of malaria, she trusted him entirely.

Brooke's story shows the power of the figure/mask crafted by Rastas for fulfilling the fantasies of, and thus forging connections with, aborofo. Brooke's story is hence a counterpart to the Rasta deployment of Rasta style to forge a new identity. But her story also illustrates the way in which Rastas were not simply out to catch a tourist—in the most basic economic sense, as mainstream Ghanaians often thought. Rather, the way Rastas sought reciprocal relations of care could generate new possibilities for living a meaningful life.

Brooke's story also provides an entry point for delimiting a final aspect of the Rasta form—emotional and immaterial labor—which, when combined with the other elements of Rastahood, produced relationships that generated mobility. If the more material aspects of Rasta practice—the production and performance of the jembe drum and the aesthetics of embodiment through Rasta style—are the obvious bases for the Rasta figure, the Rasta economy also depended on Rastas undertaking immaterial labor, such as guiding aborofo on adventurous excursions in Accra and the hinterlands, to fun hostels and hotels, and to the hippest spots, clubs, and festivals. Rastas acted as tour guides, fixers, translators, travel companions, and adventure fellows. They planned travel itineraries, gained access to locations, mediated exchanges, translated encounters, fixed problems, found doctors, nursed aborofo, and provided intimacy in the forms of friendship, instruction, flirtation, and romance. None of these tasks were contingent upon the material objects Rastas sometimes produced. Hence, even though the purchase of a jembe drum could offer the entry point into a relationship of exchange, it need not represent a point of foreclosure—the endpoint of a sale accomplished was never the goal.

Ultimately, by curating experiences that fulfilled the touristic imagination of and desire for "Africa," Rasta men acted not only as guides to Ghana itself but also, and perhaps more significantly, as guides in the exploration of a sense of self imagined in relation to a particular idea of Africanicity. In this quest, both oboroni men and women were the beneficiaries of immaterial labor on the part of Rastas, but oboroni women occupied a special place in the emotional labor of Rastahood.

This chapter, then, underscores oboroni women's perspectives of Rastahood, to bring into relief the processes and effects of Rasta labor and show how the interactions and relationships, which grew out of this labor, were both materially productive and emotionally sustaining. Highlighting both sides of the exchange—the oboroni and the Rasta, the material and the emotional—emphasizes the reciprocity entailed in these relations, which is often overlooked by both mainstream Ghanaians and scholars of tourism in the Global South. I argue throughout this book that the various forms of work that comprise Rastahood functioned to expand Rastas' sphere of agency, mobility, and opportunity and therefore, following this argument, the chapters of this book describe the incremental outward expansion of the Rasta sphere. With this chapter, in a sense, we come to Rastahood's fruition, its successful forging of connection and support, where once these were lacking for youth. In return for the love, care, and freedom Rastas proffered, oboroni women offered their own embodiment of another sort of horizon of possibility, one which could undo the immobility and waithood that Rastahood sought to resolve all along.

Romance Tourism and Rastahood as Immaterial Labor

The immaterial nature of most aspects of Rasta labor places it at the intersection of cultural and romance tourism. Therefore, this labor operates primarily within the domain of intercultural reciprocity and exchange, specifically the domain of commodified intimacy and intercultural gender relations under late capitalism. Here Rastas and aborofo find the potential for alterity to function as a mode of redemption.

Notions of immaterial labor (Hardt 1999, Hardt and Negri 2000), conceived by Michael Hardt and Antonio Negri and developed by other scholars within the informational economies of late capitalism in the Global North, provide a different way to analyze this final component of Rasta craft.[1] When Rastas acted as tour guides, translators, or cultural mediators, as they often did, they were engaging in the cognitive form of immaterial labor, which involved the production of knowledge and information, including problem-solving and brokering. Moreover, unlike the sale of a drum or the price of a drumming lesson or performance, this work was almost always unpaid, or rather, was not formally commissioned and negotiated, and Rastas rarely received direct compensation for it. Instead, Rasta immaterial labor operated in a different economic temporality, one of delayed (and hoped for) reciprocity through relation.

Examples of this kind of work abounded in Rastaworld, where one often encountered aborofo being guided or helped by Rastas in their search for something: an experience, a person, a place, or a thing. Indeed, like many ethnographies, this book would not have been written, but for the fact that Moses took on the role, for which he was never formally hired or paid, of guide, fixer, and translator for a film crew, with which I first came to Ghana. In turn, so much of what followed and constituted my fieldwork was only possible because of the immaterial labor of many others, reflecting the way Rastahood, tourism, and even, in this case, anthropology are deeply imbricated within one and another.

Ishmael, a Rasta member of Jeremiah's Northern jembe collective, told me of one example of this sort of immaterial labor. He took a young, male, American former drumming student on a three-week trip to the Upper West Region, where Ishmael was originally from. Ishmael curated the entire experience, coordinating their travel arrangements and planning their itinerary. He guided the American through Mole National Park, Ghana's largest wildlife preserve. He called on his professional contacts so that the student could see and record their musical performances. He nursed the young man when he fell ill for several days. The American covered both their expenses but did not pay Ishmael for any of the labor the trip entailed. The payment, or return, would (hopefully) come later: the young man proposed to bring Ishmael to the United States for drumming workshops and performances. It would be Ishmael's first time leaving Ghana, a more complicated, expensive mirror image of the trip to the Upper West Region. And it would be understood as one friend "hosting" another. Like all such trips, it would require a sponsorship letter, visa fees, travel expenses, room, and board while in the United States, programming, event planning, guidance, translation, and care. When Ishmael told me this story, the American had returned home, and the trip to the United States had not yet happened. I don't know if it ever did. Ishmael hoped that investing time and immaterial labor (two elements at his disposal) would produce a long-term cycle of reciprocity that would pivot the present toward a potential, desired future.

If operating primarily as a guide, fixer, and translator represents one of the principal facets of immaterial labor, in the world of the Rastas, another important facet is the affective labor of human contact, interaction, care, and the manipulation of affects—what Hardt, citing Smith, terms "the bodily mode" (Hardt 1999:96, Rutherford 2016).

Generally, affective labor has been theorized as women's work, as it most often attends carework typically gendered female, such as eldercare, childcare, marriage, domestic labor, and sexwork (Colen 1995, Federici 2020, Folbre 2012, Ginsburg and Rapp 1991, Glenn 2012, Rosenbaum 2017, Tronto 1993, Zimmerman et al. 2006).[2] In the context of this type of work, "intimate relations can be treated, understood or thought of as if they have entered the market: are bought and sold; packaged and advertised; fetishized, commercialized, or objectified; consumed or assigned values and prices; and linked in many cases to transnational mobility and migration, echoing a globalist capitalist flow of goods" (Constable 2009:50).[3] By focusing on the commodification and outsourcing (and therefore globalization) of the private, the domestic, and the intimate, the scholarship on affective labor thereby situates intimate relations within larger geopolitical and economic inequalities and, in an important corrective, turns our attention to the migrations of women rather than men and to the dramatic increase in global female mobility that defines late capitalism (see Constable 2009, England 2005 for reviews). Hence in the "new" economies, described by Hardt and others, it is mostly women, through intimate labor (as nannies, eldercare workers, nurses, domestic workers, mail-order brides, manicurists, and sex workers), who leave their home countries as migrants and send remittances back, all while mothering from afar (Constable 2016, Hondagneu-Sotelo 2001, Hondagneu-Sotelo and Avila 1997, Rosenbaum 2017, Parreñas 2001).

But the same system that drives the migration of women from the Global South to the North is also characterized by men from the Global South who *cannot* move and who are stuck in the waithood discussed throughout this book. This creates an opening for a different sort of affective labor, which takes place without migration and yet, in the case of the Rastas, also looks forward to new possibilities for mobility. This sort of affective labor arises out of the movement of "liberated," middle-class women with disposable income, in a sense comparatively wealthy counterparts to women who migrate from the Global South, but who for their part travel out of the North as tourists or for the pleasure, adventure, intimacy, and self-realization they lack at home (Kelsky 2001, Rosenbaum and Talmor 2022, Williams 2018). In this "gendered geography of power" (Pessar and Mahler 2001), it is women who travel seeking care, and men stuck at home who provide it. Brooke and Elijah's story, told above, is but one example of this geography of intimacy, which took place frequently in Rastaworld.

The oboroni women and Rasta men of Rastaworld thus fall into a category of relations termed romance tourism, a form of tourism in which sexualized embodiment is key to touristic experience, most often, but not always, in the coastal zones of the Global South.[4] This sort of tourism has received increased attention within tourism studies in recent decades, following from the acknowledgment that sex is the fourth of the "four S's of tourism 'Sun, Sand, Surf and Sex'" (Ryan and Hall 2005:ix, see Bauer and McKercher 2003) and that representations and commodifications of sex and sexualities are integral to tourism more broadly (2001:x). The connection of sex, romance, and tourism should not be surprising, since tourism is often a domain where fantasies of other places and people are made material and thereby rendered as acquirable commodities. Romance tourism can also be conceptualized as "ethnosexual" (Nagel 2000:159), when such tourism depends on the convergence between sexuality and ethnicity, on both the maintenance and crossing of boundaries of difference (hooks 1992). A consequence of such ethnosexual desire is that hosts, providing a cultural, locally specific experience, find they "have to be the bearers of very specific racialized and gendered identities" (Sanchez-Taylor 2004:48). In addition, as Susan Frohlick points out, when men are the hosts and women the guests rather than the other way around, such encounters are defined by conflicting rather than overlapping forms of power within the couple: economic, racialized power on the woman's side, and gender power on the man's. In the locales where this sort of tourism takes place, "the presence of foreign women as tourists ... profoundly influences how local men negotiate masculine identities and subjectivities" (Frohlick 2007:141). One of the responses, then, to romance touristic, ethnosexual desire, was for young men to style themselves as Rastas. Exploring the "Rasta appeal" in its country of origin, Deborah Pruitt and Suzanne LaFont (1995) proposed that, for foreign women, it was based on a combination of Rastafarian philosophy (of justice, brotherhood, and lack of prejudice), the Rastafarian emphasis on simplicity and harmony with nature, and Rasta constructions of a powerful masculinity. For local men, the image appealed for its lack of dependence upon wealth or class status. "No one expects a Rastaman to be rich" (1995:326). Instead, Rastas were expected to have new, potentially transformative forms of knowledge. As Moses once said to me, "Some people only like people from far away who can teach them new things."

The European and American women who appear in this chapter did not necessarily travel with romance tourism in mind. Instead, they first traveled to Ghana for varying reasons, as cultural tourists, foreign students, or volunteers in their early 20s. But they quickly formed part of a larger group of young aborofo travelers who inhabited Rastaworld. It could, nonetheless, be within days of arriving in Ghana that these women met Rastas and, soon after, were in romantic relations with them. Among these women, there were certainly those who fulfilled the stereotype envisaged by the phrase "to catch a tourist"—oboroni girls, who hooked up with a Rasta youth in a short-term, purely sexual way, and Rastas, as well as other working-class men, who wanted to be with an oboroni woman only to get to aborokyire. But the people who tell their stories in this book sought and had longer, more multivalent relationships. In particular, the women of this chapter remained connected to Ghana for periods as long as three to ten years after establishing romantic relationships with Ghanaian men. Consequently, my interviews with these women, like my interviews with Rastas, took place repeatedly over long periods. They are reflections from different points in time, from within very different chapters, in their relationships with both the Rastas and Ghana.

Thus, while the rich literatures on romance tourism, immaterial labor, and global carework are useful for understanding the demand women bring to Rastaworld, the labor men perform, and economic inequity of this exchange, long-term ethnography can also move beyond these frameworks, exploring reciprocity-across-difference and reciprocity-over-time (see Cole and Thomas 2009). With this alternative framework in mind, I draw on Bianca Williams's concept of "emotional transnationalism" (Williams 2018, Wolf 2002) and, especially, the concept of a particular sort of freedom (Berlin 1971, McWhorter 2013, Turner 1974) that many of the men and women, who inhabited the contact zone, realized through a sincere shared experience.

Theorizing Freedom

> The word that I would think about was the freedom. I think Rasta is an invitation for the foreigner to be free, to lose their inhibitions. Maybe they didn't feel sexy before, or their body wasn't respected or worshipped back home, so they feel a freedom in their body, a freedom to move, a freedom to enter into this laid back, alternative world or alternative reality, where you feel sort of like a queen, kind of worshipped. It's even in how Rastas talk, "Oh, my queen." It's entry into a world and places that you normally wouldn't get to go and see. You feel good when you're a part of that, whether you're smoking or drumming or dancing. It's an entry into this sort of feel-good, let-it-all-loose world. (Angel, an oboroni woman)

For Angel, the concept of the "freedom" women experienced in Rastaworld had several facets. It was a "freedom-from" (Berlin [1958] 1971, see below) the experience of their embodied selves at home not only as inhibited but also as unlovable and as disrespected. Hence, "Rasta," (Angel uses the word as if to describe an essence or energy) in contrast, functioned as "an invitation," a freedom-to "move," to be "in their body," while simultaneously producing "an entry" into an "alternative reality," where a woman is "worshipped," where you are addressed as "queen" and might feel respected. Even more, Rasta offered an "entry" into "a world and places that you normally wouldn't get to go and see." In this world, crucially generated through Rastas-as-men and Africa-as-place, romance, adventure, cultural immersion, smoking, drumming, and dancing are all conduits toward a generalized sense of feeling good, letting it all loose—being free.

For their part, Rastas understood what oboroni women were experiencing. As Noah explained:

> Rasta means freedom; Rasta means righteousness. When a foreigner comes to us, they feel safe, especially women. … We know that you are tourists. You don't know anybody [here], so we approach you nicely. So when they come to Africa and they see us, they feel that openness, they feel that ease…. They don't know us. We would ask ourselves, "Why is it that this person doesn't know us, and [yet] they trust us?" It doesn't even take a minute, and

the first thing they will say is, "I like your dread." It was something beyond our understanding, but we began to understand that our dread says something about us. …People feel free to hang out with us, to hook up with us, especially girls. The dread attracts, and mostly attracts women. And chances are, when you are in America, you have trouble with your relationship, and then in Africa, you have this nice guy who is always there for you. When you come to Africa, we give you more time. We respect you. You don't know the place; we show you around.[5]

No matter what it became later, a relationship between an oboroni woman and a Rasta man originated in an exchange of images (or figures). The Rasta figure—"the dread"—instantly conveyed something to women, which made them feel trust and ease, on the one hand, and openness and attraction on the other. The figure was a promise: here was the fantasy made flesh, "a nice guy who is always there for you," who gives you time, attention, and respect, and who, at the same time, was a portal into the alternate wonder that was "Africa." Together, these produced what both Angel and Noah describe as "freedom." Similarly, for Rastas, the oboroni woman figure was also a promise and a fantasy made flesh.

These invocations of Rasta and Rastaworld as foci for experiences of freedom raise questions about the different kinds of freedom women do and do not experience. Feminist thinkers have asked these questions by taking up Isaiah Berlin's notions of freedom-from and freedom-to, or negative versus positive liberty (Berlin [1958] 1971). Negative liberty consists of an absence of external constraints, imposed by an external "other," while positive liberty is the absence of internal barriers. For women, both types of freedom usually get formulated within the context of patriarchy, which arguably produces externally generated constraints but presents and pathologizes them as inner barriers.

Angel uses the word "free," above, in both ways: "to be free [from] … inhibitions" and to "feel a freedom in [one's] body, a freedom to move, a freedom to enter into this laid back, alternative world or alternative reality …." In addition, the "from," for Angel, refers to constraints back home and the "to" refers to freedoms found elsewhere, in Africa. In this way, Angel, and other subjects of romance tourism, creates a geography of spatialized freedom, around which home and Africa find themselves in opposition.[6] This Africa, however, was not the one of mainstream "Christianity," as Brooke explains, which, in her view, imposed its own, new constraints upon women.

Angel's words also confirmed the conflation of Rasta men, Rasta style, and Ghanaian/African Rastaworld as sites of freedom. Her description of a certain "alternative world or alternative reality" distilled a by-now-familiar, highly durable Western imagining of Africa, as a site for the transformation of the self, wherein one critiques and questions Western life, through consumption of, immersion in, and contact with difference. This is a common aspect of the touristic experience, but it has been overdetermined in Africa due to enduring racialized and colonial tropes. Since it was impossible to interact with "Africa" per se, one achieved what was imagined to be the same thing by interacting with an individual who seemed to embody Africanicity or consuming a cultural practice which was understood to materialize the very same. For oboroni women, then, Rasta affective labor opened a portal to a back region (Goffman 1959) not only of Ghana/Africa but also of the self. Rastaworld—wherever it happened to manifest—offered a site of freedom, of discovery, and of self-discovery, often through the medium of embodied culture (see Bizas 2014, Faier 2007, Kelsky 2001, Sawyer 2006, Williams 2018).

Alongside feminist scholars, Richard Turner draws on Berlin's concept of freedom to explain the experience of tourism more generally (and regardless of gender) as a series of "liminoid" situations of leisure-time defined by "freedom-from" the rigid temporal and behavioral constraints of work and life in (post)industrial societies, and "freedom-to" transform through the accrual of new forms of knowledge and experience (Turner 1974:68). For Turner, such forms of tourism combine *communitas*, play, and the ethical consumption of culture into a modern analogue of religious pilgrimage, a quest for an alternative to the shallowness and alienation of daily life (Turner 1974). To use Erving Goffman's dramaturgical metaphor of the back region, such tourist-pilgrims are seeking a deep experience of reality and solidarity lacking in their own lives, through ritual journeys into the backstages and real lives of other time-spaces, exchanging the ordinary for the ecstatic and the different. However, if Turner does not intend his analysis of tourism to specifically address the gendered nature of freedom and unfreedom, it is nonetheless possible to imagine why tourism, as a path out of the alienation of one's daily existence, may have an especially powerful and gendered appeal to women, who are accustomed to the significant constraints of patriarchy.

But if tourists, and especially oboroni women, sought, amongst the Rastas, genuine connection, Rastas themselves also sometimes longed for the same depth. Along these lines, a Rasta named Josiah once told me, "Very few people—Black or White, men or women—are righteous. African girls will go after you for money, but White women will come just for fun." He explained that a lot of the Rastas were not in love with their foreign girlfriends, and a lot of White girls took their relationships with Rasta men too lightly. He wanted to be with someone he loved, he didn't want to get hurt, and he didn't want to fit the stereotype, projected by Ghanaians and oboroni alike, that because he was Rasta, he must be interested in oboroni women. Josiah's point of view resonates with the ways in which the oboroni fantasy of freedom was constructed in Rastaworld: it was yet another version of the tacit, secularist, Western assumption that "flaunted sexuality is a token if not a measure of women's freedom and equality" (Abu-Lughod 2015, Brown 2012:52, Deveaux 2018). Living out this version of freedom in Ghana was an exercise of racial, geographic, and economic privilege, even as women experienced it as an escape from patriarchal power.

Points of Entry

As mentioned previously, Rastas occupied the touristic (and young expatriate) contact zone, going to the places where oboroni women hung out and "welcoming them" to Ghana. They did so in order to meet oboroni women where they were, literally and, as Brooke's story demonstrates, emotionally. To a degree, Rastas produced freedom for women simply by being present, present upon their arrival in Ghana, a friendly face in an unfamiliar place, and present to respond to the complex needs these women brought with them, in search of new possibilities for resolution.

"They were at the places where White girls go—like the Arts Centre, the beach, and Akuma Village," Rose, another oboroni woman I met in Accra, recounted.[7] Rose herself met a Rasta musician, Abilene, at Labadi Beach, an established Rasta beach in Accra, where reggae was regularly performed.[8] Labadi was also one of the most welcoming, cosmopolitan places in early 2000s Accra, the crowd—laying out, swimming, strolling, and enjoying beers in the shade of sun umbrellas—included both working class and elite Ghanaians, expats, tourists, and everyone in between. Among them, child acrobats, hawkers, and food sellers plied their trades. People on horseback wove their way between beach football games. Abilene and his friends played football there every Sunday. Afterward, they would circulate through the rows of wooden lounge chairs, selling art items, playing music, and asking for donations.

Brooke had met Elijah at a place similarly frequented by aborofo and others, the Shrine, situated behind the School of Performing Arts at the University, where most foreign students took classes in drumming, dancing, and Ghanaian performing arts. Other couples met at Alliance Française, which hosted a popular Wednesday night live music performance that also drew a mixed crowd. And of course, many couples met at the Arts Centre or Akuma Village, a Rasta village, oboroni hostel, and club-restaurant-performance venue.

These were the places where White girls—and Rastas—went, where Rastas spent time as tourists wanted to spend time. "If you're on vacation, and you're partying and waking up at 1 p.m. wanting to party some more, you aren't going to be connecting with the 9-to-5 people, you're going to seek out the Rastas," Rose explained. Noah held a similar view of the shared spatiotemporality of Rastas and oboroni girls. To him, oboroni girls, like Rastas, were "free."

> When I was in school, I had a Ghanaian girlfriend, but Ghanaian parents are strict. My Ghanaian girlfriend was so restricted. She was not allowed to go out. [But as a man] I had freedom and foreign girls also had that freedom that I had. They were also able to go to clubs and party and drink. [So] they find another person that is like them. We just link up and hook up like that.

This way of spending time and occupying space set both oboroni women and Rasta men apart from mainstream Ghanaian society. "Rastas are outsiders in Ghanaian society and White girls are outsiders in Ghanaian society, and what do outsiders do? They flock to each other," Rose explained. "And for the most part with Rasta style, there is an expectation that people will be more friendly and open to talking to you, not because they want something from you because of your touristness, but because they identify themselves as outsiders, like you."

Rose and Noah's words elucidate how Rasta-oboroni relations took place in and reproduced an intercultural space—a contact zone that differed from both its adjacent territories in terms of a certain sense of "freedom." In this case, for both Rasta and oboroni women, it was freedom—from the societal constraints of home and freedom—to transform oneself and expand one's sphere of agency. And it carried an emotional valence. As Brooke said of Elijah, "It was just as exciting for him."

But there was also an emotional valence for women, within this space. It was the affective style evinced by Rastas, which became possible away from other societal constraints, a combination of care, acknowledgment, ease, and pleasure. Rose explains, "It was more the smile, right? There was a big smile, and he was really open, and he wasn't grabby—he wasn't trying to kiss me or sex me—he was just chill."

Rasta affect made these oboroni women feel that they were authentically desired, in contrast to how certain other Ghanaian men made them feel, as though they were desired only for the financial and visa access they could provide to get out of Ghana. Rasta desire felt genuine. It was located in a place where oboroni women and Rastas came together to be present together, not in an orientation toward somewhere else. Rastas weren't out "to catch a tourist."

This affect of containment (relative to the ways other Ghanaian men interacted with oboroni women) was also accompanied by an openness, a vulnerable honesty, and a lack of judgment (relative to men back home). "I'm gonna say they made an effort," Rose explained. Jane, an American woman, corroborated:

Ghanaian men are *so* charming. They say all the things that you want your American boyfriend to say, [the things] you know that they're thinking ... but they'll never actually say it to you ... [Ghanaian men] are very open with their feelings and much more in touch with their initial impulsive feelings than Americans. They're a lot more intense [and] they fall harder. ... And they're very openminded about everything like bodies. I don't ever feel self-conscious about my body with Ghanaian men. I feel like they just like who you are and they're not so concerned about, "Oh, what are people going to think about me being with this [woman]." ... There are so many more complications to being with guys at home. ... [Here] they just feel a lot more comfortable with themselves as people.

This emotional and sexual openness accompanied a political and social one. "In some ways, as an oboroni, you expect someone with Rasta style to be more progressive and social justice-minded and all of those things that go along with arts and creativity and freedom of expression," Rose explained.

Thus, for aborofo, Rasta style signaled a compelling combination of traits: an emotional honesty, a progressive political orientation, artistic creativity and freedom of expression, and a friendly, curious openness to the world, combined with an attachment to, pride in, and knowledge of Africa and its traditional culture. Altogether, this created a space in which oboroni women and Rastas moved together, feeling more free. It was always present, found even in the way Rastas moved through the city. "He liked to walk the back routes everywhere; he liked to explore," Rose said of Abilene. "He would never be walking the main road."

Finally, for these women, Rastas gave them access to adventures in the less touristic (read poor) backstages of Ghanaian life. Rastas themselves recognized that aborofo sought to connect with them because their lower class status provided access to a more "authentic" Ghana. As Josiah explained,

> Mostly when you travel and you want to see a country well, it's the local people, the simple people, who are going to introduce you to the best place ever. You cannot go to a lawyer or a doctor—he will snob you. But the local guy will take you anywhere. That is why they [young Ghanaian men] capitalize [on this touristic desire] and turn Rasta. They are all from the pure, typical ghettos: Jamestown, Adabraka.

Poverty, then, was key to authenticity; experiencing it was key to self-transformation. "Local" connoted genuineness; ghettos were "pure, typical." Unlike middle- and upper class mainstream Ghanaians, underprivileged young men with no formal education were the ones who knew the country well, would give you the time, and could take you to the best places ever.

The presuppositions and misconceptions that shaped this "poor but happy" stereotype (Ferguson 2006, Simpson 2004, Swan 2012) were part of the touristic oboroni fantasy that Rastaworld fulfilled. But for those women who spent years in Ghana, the connection to their Rasta partners eventually led to an understanding of Ghana's complexity and of their own racialized associations of Africa with poverty. As Rose explained, looking back:

> No one ever had a Rasta boyfriend who was living in suburbia. Rasta guys were living very minimalist. And it was always that thing among the White girls of who could be more minimalist and roots and authentic. I remember once going home with this Australian girl and her Rasta boyfriend, and they were living in this place that was half concrete bricks and half nothing. I was like, "You win, you for-all-time win, you are the most, you win." You want the "authentic" experience because you've been fed something about what authentic is, and you don't realize that sharing food with a middle-class family can be authentic. The authenticity is about poverty, because Africa is poor, and you've been fed all that bullshit [about how] the people have so little but they are so happy.

Ultimately, for many of these women, Rastahood iterated the mythic Africa of the Western imagination, an Africa that conflated race and class, Blackness with poverty, and Whiteness with wealth.

Whiteness Is Wealth: The Other Horizon,
a Different Kind of Freedom

This association of poverty with Blackness (and thus Rastahood) and of Whiteness with wealth were essential to the initial connection between Rasta men and oboroni women. In the space of the contact zone (at the beach, dancing, taking the back routes), both men and women not only moved together, but they also metonymically embodied for each other an imagined place with transformative potential, a freedom from restraints they experienced at home, and a new horizon of possibility.[9] Just as Rastas embodied Africa and the solution it promised, oboroni women embodied aborokyire, a "White, non-African elsewhere" (Ferguson 2006:151) and provided a solution to the economic quandary and problem of masculinity that had produced Rastahood in the first place.

Becoming Rasta was almost always a pivot outward from within a Ghanaian socio-economic system in which youth were trapped in poverty, in economic, temporal, and spatial immobility, unable to support dependents and be men. The predominantly White, politically progressive, feminist women, who found Rasta youth desirable despite (or arguably because of) their poverty, willingly covered the expenses of living in Rastaworld and traveling beyond it. Unlike most Ghanaians and many foreigners, they did not perceive Rastas as freeloaders or hustlers out to catch a tourist. They saw them as artists—people who worked hard, though not in established professions. And they were drawn to the freedom that seemed to be embodied in these choices. Oboroni women, hence, viewed Rasta style itself as a manifestation of Rasta men's independence, creativity, and joy. Rose explained:

> All of those guys had a hustle, [but] I don't mean a hustle of tourists, I mean a hustle for themselves. It's not like any of those guys had a 9-to-5 job. I really don't think that the majority were specifically doing it to catch a White girl. I would say [only] 15 percent were specifically doing it because [they thought], "If I wear this style, White girls will be interested in me." I think for most of them it was a side perk. There was a joy in that style, a joy in the adornment, the necklaces, the "I'm gonna get up this morning and dress this way because I like it, for the pure aesthetics of it."

"There was a joy." Oboroni women wanted to be with these Rasta men, for everything they represented and all the ways they received oboroni women that was so unlike their experiences at home. They valued Rastas for who they were, how they moved through the world, and how they treated others. And thus, given their own relative economic advantages, they were willing to support Rasta men in order to be together, to support their artistic pursuits, and ultimately help them achieve economic and existential transformation.

From the perspective of Rasta men, this was quite different than what they might experience with Ghanaian women. "You need a wife who will work with you, who will put her mind with yours to build something, not a queen who will sit and ask you for everything," Ras Caleb explained to me once, as we sat out the rain at Akwaaba Restaurant at the Arts Centre. By "queen," he meant "African queen," Rastas' descriptor for Ghanaian women who expected, and in many senses needed, men to support them financially and who thus never considered Rastas as potential life partners.

Noah elaborated:

> Don't forget you guys are rich. You were coming from a rich country, so you have more money than us, so who wouldn't like to hook up with someone who has money? If you have a Ghanaian girlfriend, you have to do everything for her. [But] when you guys come, you understand everything. You understand these guys are nice, they're just not rich. Because of this, the foreign girls were more attractive to us. And to be honest they were more humble to us....So it was a change. Everyone wanted to have a White girl; they were cool. You were respected [because] she is White [and] she comes from a foreign country.

"Nice...just not rich." As Pruitt and LaFont (1995) discovered in the Jamaican context, Rastahood permitted men to construct a desirable masculinity not based on wealth. Rastas knew that their female counterparts in Ghana—young women from similar economically struggling families—were just as disadvantaged as they were. This is in part why Noah became a hustler, to put his sister through school; he feared the alternative of her having to find a boyfriend to support her. But most Rastas could not aid women in this way; instead needing oboroni women to help them. And unlike "African queens," oboroni women were "humble": they did not judge or reject Rastas for their inability to achieve adult/proper masculinity. Instead, they "understood everything"—their self-awareness of their economic privilege, feminist views on gender relations, and liberal political stances on the global distribution of capital allowed Rastas to solve their domestic issues.

Hence, many oboroni women supported Rastas' projects financially, helping them transform into men worthy of respect. Jeremiah, Samson, Gideon—Rastas who had their own drum shops—all received an initial loan or gift from an oboroni woman to get them started. In other cases, several women that I knew traveled back and forth between Ghana and their home countries, working at home to bring savings to Ghana to invest in businesses started by their partners. Still others settled in Ghana and worked in the expatriate sector—in embassies, the UN, NGOs, foreign businesses, and the international school—earning dollars, euros, etc., and supporting their partners as they worked on their dreams. Brooke had done exactly this. A few years into their relationship, she and Elijah bought land and started a small ecotourism venture, investing the expatriate salary she earned into the ecofarm and fulfilling Elijah's dream. Rose, working a similar job, supported Abilene's musical career, lending him money even after they separated.

But there were also other things Rastas got out of their relationships with oboroni women. As Noah said, "everyone wanted to have a White girl," "they were cool." A "White girl" was desirable not just for the economic support she provided and the refreshing desire she exhibited but also for the respect she evoked as an accoutrement to Rasta masculine self-presentation. "You were respected [because] she is White [and] she comes from a foreign country." As Jemima Pierre states in her study of racialization in Ghana, Whiteness in Ghana, as elsewhere, transcends the corporeal; it "articulates with racialized-as-White bodies, all the while moving beyond such bodies and expressing itself in other representations of itself—such as culture, aesthetics, wealth, and so on" (Pierre 2012:72). Even more, in the Ghanaian context, Whiteness represented "development, modernity, intelligence, innovation, technology, cultural and aesthetic superiority, and economic and political domination … all things that denote value in today's world" (2012:74).[10] As discussed in the introduction, the term oboroni, while associated with Whiteness, is also used to refer to Asians, Latin Americans, diasporic Blacks, and even to Ghanaians returning from abroad who act White (Pierre 2012:77). As Pierre explains,

> To be considered obruni, even if jokingly, is to be associated with the class and cultural standing of Whites (and Whiteness) in Ghana. Significantly, because the term is also employed in describing seemingly rich and culturally different Ghanaians and other Blacks, it signals a clear association made between Whiteness as racialized identity and Whiteness as a particular class status, cultural standing, education level, and outlook.
>
> (2012:77)

Citing Franz Fanon's "you are rich because you are White, you are White because you are rich" (Fanon 1967:10), and Cheryl Harris's foundational formulation of "whiteness as property" (1993), Pierre points to "the ways that Whiteness assumes privilege and is deployed as identity, status, and property" (Pierre 2012:77). Thus, oboroniness transcended corporeal race, even as almost all of the oboroni women whom I knew to be romantically involved with Rastas were White, and none were Black.

The association of Whiteness with wealth and status meant that oboroni women as accoutrements did evoke as much respect as they could have. On the other hand, they often disappointed by not *acting* White, or rather, for having a particular White style that included a hippy-in-Africa fashion (rather than more "correct" clothing). They also refused to conspicuously display or perform their wealth, by engaging in practices such using public transportation and eating street food in ways that elite Ghanaians and other foreigners would not (see Pierre 2012:79–82). Hence although Rastas enjoyed oboroni women's corporeal Whiteness, they likely would have also benefitted from women wearing, rather than discarding, other, noncorporeal elements of Whiteness clearly linked to wealth and class status. But it was exactly these aspects of Whiteness that women sought to leave behind and refused to deploy in Ghana. This rejection too, as Rose eventually understood, was a privilege of Whiteness, just as the equation of African authenticity with poverty and nonmodernity was a product of the racial formation within which oboroni women and Rastas operated.

Beyond economic support and the unstable benefits of racial signification, oboroni women most importantly provided a "spatial fix" (Mains 2007) for Rasta immobility. They supported Rastas in the project of building a shared life abroad.

Just as Rastas provided these women entry into Ghana, freeing them from the constraints of home, women freed Rasta men from the never-ending circularity of the hustle, of waithood and precarity, replacing it with a forward, upward motion into a "better future." This took multiple forms: women traveled ahead and saved money to pay for visa fees (including the expensive fiancé visa), for travel, and to support men while they were legally forbidden to work abroad or unable to do so due to language barriers. As Rastas had done for them in Ghana, in their home countries, women worked as translators, fixers, and mediators.

In the end, for both parties, it was a reciprocal mode of relation, even within the complicated framework of inequality. Noah himself summed up the nature of this reciprocity:

> They … helped us by getting visas to come to them … No one wants to live the sort of life that I was living in Ghana, because I was always hustling. I didn't have stand. If you were my girl and I married you and came to America, you establish me, [now] I have a firm stand. You guys helped us in [giving us] a way [to] a better future. And we also helped you by coming into your lives, bringing you happiness when you were with us in Africa. If you look at the number of African guys who have married White girls, we are so many. It's vice versa and it helps.

As Noah portrays it, the reciprocity and transformative potential of Rasta-oboroni relations are a primary feature ("it's vice versa and it helps"). This is reflected in the way that both Africa and aborokyire became geographic manifestations of a desired alternate reality unavailable at home, the way that Rastaworld acted as a portal that opened in both directions. For Noah, both men and women got what they needed from each other in the land of the other. Oboroni women found "happiness" and Rasta men gained a firm "stand." Along the same lines, Brooke explained, women also viewed the exchange as reciprocal, for they understood the economic realities that both they and Rastas had inherited:

> It never felt like freeloading because he would give everything that he had to give. Everything that he could access, he would access. But [in] the end, he did not know anybody with money …. Even in retrospect, I never felt like I was taken advantage of for that. That was obvious; that was a given.

From Tourism to Emotional Transnationalism

Rasta-oboroni relations, nonetheless, were not necessarily an end in themselves. Rather, as we have seen, they were sites of self-transformation and reciprocal care, for persons alienated, socio-economically or affectively, from their places of origin. It was from this perspective that Noah told me in London in 2016, "A White woman brought me here." This was years after the relationship had ended, after he had ceased to be a Rasta. And yet, when Noah looked back, he recalled, "She helped me, and I will always be grateful to her."

Many of the Rasta men who appear in this book were helped by an oboroni woman and helped an oboroni woman in turn. Their search for fulfillment and discovery both drew upon and worked against cultural gendered scripts, or schema, that delimit what we must do to be happy, natural, and good (Ahmed 2010, Johnson-Hanks 2007). Sarah Ahmed suggests this struggle over happiness forms a political horizon, around which operates a set of enforced adjustments and behind which lies the loss "of other possible ways of living" (Ahmed 2010:59). In contrast, for Ahmed, "happy objects," "point us toward happiness" (2010:27), make us "move in a direction toward somewhere *else*, somewhere *after*" (2010:26).

Encountering one another within the contact zone of Rastaworld, then, presented Rasta men and oboroni women with potential happy objects and an accompanying way beyond the limits of inherited gender scripts. For oboroni women, Rastahood was a happy object full of affective potential. As Noah said, the dread elicited both desire and trust, made women feel both safe and excited (see Nagel 2003:207). It held a cluster of promises (Berlant 2011) about freedom, joy, and authentic presence. Similarly, for Rasta men, oboroni women's Whiteness was a happy object, which itself bore a promise of freedom, support, respect, and economic agency. In both cases, happiness derived from a promise of motion in the direction of a spatiotemporal elsewhere—a land beyond the known horizon (aborokyire).

This experience of self-transformation through desire by and for the other reinscribes touristic experience into what Bianca C. Williams, building from Diane Wolf, calls emotional transnationalism. Describing the touristic experience of female, African-American "Jamaicaholics," Williams asks: "Why do people seek out diasporic and transnational experience?" "How does this desire reflect nationally specific affective and political economies of race and gender?" (Williams 2018:5). For the women Williams works with, Jamaica provides freedom—from American racism and sexism, from lives working and caring for others and freedom—to find leisure, belonging, and their own self-care. In this way, Williams connects their emotional lives to transnational mobility.

In a similar vein, for oboroni women and Rastas, Africa and an aborokyire hold the promise of emotional transnationalism. The shift in sociocultural context that Rastas and oboroni women bring to each other, in the contact zone, allows varying dynamics of power and privilege to coexist with and permeate practices of reciprocity and solidarity (Williams 2018:6). Indeed, this duality of power and reciprocity defines Rasta-oboroni relations, which always took as a given the imbalance of money and mobility within the couple. As Zelizer (2005) and others have argued, the notion that the world splits sharply into separate spheres of rationality and sentiment, of economic self-interest and authentic emotion, of "true love versus material motivations" usually falls short of actual experience (Faier 2007, Mauss 1967, Weiner 1992, Zelizer 2005:13). Whereas when Rastas and oboroni women used words like freedom, respect, and help, when they spoke in a language of reciprocity, they offered a more nuanced way of thinking about intimacy as shot through and shaped by unequal power relations, even while being aware of the transactional stereotype they were enacting.

Hence, even as oboroni women objectified and idealized Rastas, and Rastas did the same, they also moved beyond their initial reception of each other to a place of mutual care and then to a grasp of each other's realities. Rastas thus might at first read oboroni women's bodies, especially their hypervisible Whiteness, within the Ghanaian gaze, as signs of access and possibility. And oboroni women might at first read "the dread" and the hypervisible Blackness and Africanness of Rastas, within the oboroni gaze, as dewy-eyed signs of freedom, inhibition, and uncomplicated acceptance and joy. They thus might each first approach the other in terms of an exchange of images, of fantasies about what the other represents. But they also, over time, came to be attuned to each other and the material reality of each other's lives.

When Rastas read oboroni women's needs and then embodied their fulfillment, they helped women feel "free" to be fully themselves, to experience themselves as full persons, in a way that they could not at home. Rastas did this not for direct payment but with the hope of delayed, reciprocal care. Oboroni care took a different form—it was incorporated into the sense that women's bodies were symbols of status for Rasta men, which gave them a respect they lacked, and more importantly, oboroni women "worked with" Rasta men, supporting their entrepreneurship, helping them travel, and enabling their life projects as men.

Hence, Rastas and oboroni women each knew in their own way and their own time that they represented something to the other, something that could be used for a purpose. Rastas knew that they could take control of the image of Africa that was so appealing to aborofo and make it work for them. Similarly, oboroni women knew—and were endlessly told—that Rastas were only after them for money or a ticket out of Ghana and that they appealed to Rastas, through their Whiteness, as an embodiment of status. And over time, these women also came to understand the reality of Ghana and the colonial imagery they had imposed upon it. And yet, beyond all this, there was always a surplus of affect and individuality that came through in both men's and women's accounts of their time and the lives they constructed together. In a more profound and more complicated and more realistic sense, there was freedom.

Notes

1 In Hardt's study of the postmodernization of the labor economy, the concept of immaterial labor speaks to the "change in the quality of labor and the nature of laboring processes, whose end product is no longer a material or durable good but a service, information, communication, knowledge and/or the production and manipulation of affects" (Hardt 1999:93).

2 Arlie Hochschild's groundbreaking work on the emotional labor of producing affect in oneself so as to produce affect in the other demonstrating how the production of affects in the recipient depended on the control of affects in one's client is of special value (1983).

3 But see the work of Viviana Zelizer for the broader interpenetration of intimacy and economic activity in social life, in everything from marriage to inheritance (2005).

4 Much of the early scholarship on romance tourism took the Caribbean as its context (Albuquerque 1998, Brennan 2004, Frohlick 2007, Jeffreys 2003, Kempadoo 1999, Mullings 2000, O'Connell Davidson and Sánchez-Taylor 2005, Pruitt and LaFont 1995, Sánchez-Taylor 2001, 2004, Sharpe and Pinto 2006), but Ecuador (Meisch 1995), Egypt

(Jacobs 2009), Indonesia (Dahles and Bras 1999), and Nepal (Yamaga 2006) have also been discussed. In Africa, the Gambia (Ebere and Charles-Ebere 2018, Ebron 1997, 2002, McCombes 2012, Milou 2010, Nyanzi et al. 2007, Wagner and Yamba 1986, Williams 2018) and Kenya (Kasfir 1999, 2004, Kibicho 2016, Meiu 2009, 2017) have received the most attention. Padilla et al.'s edited volume (2007) demonstrates the range of this literature.

5 Noah's elision of "you" and "they" in talking to me reflects my positionality. I was "you," a representative of oboroni womanhood and a "tourist" (a word he uses to more broadly encompass several kinds of travelers). But I also wasn't "you": I had not become romantically involved with a Rasta. I knew them, and they me, in a different way.

6 While falling outside the scope of this chapter, it is important to complicate the term "liberated" when discussing middle-class women with disposable income who engage in global travel. As Alexander and Mohanty state, "capitalism is […] a set of processes mediated through the simultaneous operation of gendered, sexualized, and racial hierarchies," and women's agency and desire are always negotiated in the context of traditional gender ideologies and inequalities (Alexander and Mohanty 2012: xxi). Thus, an intersectional analysis is needed to understand the different forms of power—economic, geographic, ideological, gendered, and racialized—that men and women bring to an encounter and to move beyond essentialized, unsituated gender categories (Alexander and Mohanty 2012, Bernstein 2007, Faier 2007). See Ebron (2002), Frohlick (2007), Ray (2015), Stoler (2002), Williams (2018), and Kelsky (2001) for such analyses of specific cases, including Ghana (Ray) and Gambia (Ebron) in West Africa.

7 Like many of the women quoted in this chapter, Rose's relationship with a Rasta man happened almost immediately upon arrival in Ghana for what was intended as a year of volunteer work as a part of the British VSO (Volunteer Service Overseas). The relationship led to an unplanned pregnancy, and although Rose and Abiliene separated soon after the birth of their child, Rose stayed in Ghana for seven years as a result. Thus, although she was not a tourist, her initial connection with Rastaworld happened within a similar time frame, when Ghana was an unknown entity and Rastahood embodied it compellingly.

8 Partly due to Rocky Dawuni and Carrie Sullivan (whose interviews appear in the previous chapter), who created *Independent Splash*, a popular reggae festival that has taken place in Labadi intermittently since 2001.

9 I take inspiration here from Thomas Abercrombie's description of himself as a "walking frontier" during his fieldwork in Bolivia (Abercrombie 1998:51).

10 For seminal definitions of Whiteness, see Harris (1993), Dyer (1997), and Frankenberg (1997). I am also deeply informed by work on Whiteness and womanhood, including Frankenberg (1993) and Twine (2010), as well as work on race and gender in colonial contexts (McClintock 1995, Stoller 1995, 2002, and Ray 2015).

Conclusion

In the Beckoning Elsewhere

The first time I traveled to Denmark in 2012, I landed in Rastaworld. Moses picked me up at the airport with a close, female friend—an "adopted oboroni," in her words—who had been going to Ghana regularly since her first stint there as a volunteer in 1999, and whose life in Copenhagen was as intercultural as Rastaworld in Ghana was, organized around drumming and dancing filled with Ghanaian men, Danish women, and their children. Now, Moses was crashing on her couch, as he was in between apartments, in between jobs, and planning a trip to Ghana. We went to a bar, where I was astounded to see Nimrod, a Rasta from the Arts Centre I hadn't seen in years, DJing. Turns out he had been living in Copenhagen for nearly ten years.

In the car, Moses, who had seen that I had brought my camera, asked if I could film three performances he had set up for the week. An Argentinian guy wanted to bring him to Buenos Aires to run some workshops and needed a demo to get funders for the trip. Moses said, "Ruti, me these days, I am open to everything, because you never know. They say when one door closes, another door opens."

We went to the house of Caleb, one of the original members of African Drumbeat, who had been among the first to arrive in Denmark and was now well established, working a government job and playing in a Danish-African reggae band on the side. His apartment, where he lived alone when his two children—whom he shared with his Danish ex-wife—weren't staying with him, was a Rasta gathering place. Moses, an excellent cook, had made *red red*, my favorite Ghanaian dish, in honor of my coming.

People were there whom I hadn't seen in years, among them Simeon, whom I had met my first summer in Ghana in 2001, while working on a documentary film crew, and never seen again. In fact, he had been my first Rasta acquaintance, a dancer from Cape Coast who was also the guide-fixer-mediator for the people who had hired us to make the film. It was he who had introduced me to Moses, who in turn had introduced me to the Arts Centre. Over whiskey, we covered everything from that first summer to the present day. He had first come to Denmark with the Ghanaian-Danish girlfriend I had met that summer.

"And then?" I asked.

The pressures of moving to Denmark had split them up. "For the first six months, you are not allowed to work. You are in the house like a prisoner of circumstance," he explained. Eleven years later, he was working in management, happily married to a different woman, and had a child.

DOI: 10.4324/9780429244568-7

"Tell me the next time you are going to Ghana," he said. "Me, I don't go to Ghana just for myself." He told me about his last trip the previous summer, and how he always spent so much money when he was there (a common complaint).

Two weeks after, he returned to Denmark, his father had died, and "people didn't think I would make it back." But he was good at saving money, so he came back and stayed for a month. "If it was me," he said, the funeral would have been for immediate family, but his mother wanted it "traditional."

"It would have made your father proud," I said.

"Yes," he responded. "Sometimes you just get one Fanta and some chips and that's it," he said of the fare offered to funeral guests at some of the notoriously large Ghanaian funerals, but he had paid for an open bar and a full meal.

Alone with me in the kitchen, Caleb said I should talk to Moses. He was resisting settling down into a steady job. "You cannot hustle in Europe," Caleb said. "It's not like back home, where you can always find houses to feed you. Here, everyone has their own problems." People had helped him before, but they were getting tired. "He is lucky. When we came, there weren't so many people already here."

Moses had told me his version of this. He was struggling to make it as a drummer, and that was all he had ever been or wanted to be. In Ghana, even though he struggled financially, Moses could always play. He hadn't imagined it would be so difficult in Denmark.

"Back home in Africa," he said:

> We don't learn the music from school, we learn it by family. But here, because they don't have the rhythm, they learn it by notes. But the bad thing is, we who have it, we can't make it, you know? And of course, we the musicians have to come together and show the world that we can make it. But each and everybody has their own way of thinking. A lot of people are working; they're not making music anymore. It's bad. It's music that brought you here. You should do both. They leave their culture behind, because they don't want to fight for it. Why not fight for it, rather than leaving it somewhere, or giving it to someone else?

In this complex quote, Moses named several obstacles to success as a musician in Europe. First was the difference between oral and written modes of musical instruction. Moses had never learned musical notation, and without it, it was nearly impossible to land teaching gigs or to play in most bands. But then Moses shifts his criticism to other Ghanaian (and African) musicians, who are not sticking together, sticking to music and African culture, instead moving on, "leaving it somewhere, or giving it to someone else." In Ghana, they had all worked together, using "their culture" to get ahead. Now that they had arrived, he saw many people leaving it behind.

The six or seven people in the apartment that evening were in a range of relations to Denmark. Some, like Caleb and Simeon, were firmly established there, made a good living in jobs unrelated to African culture, but remained connected to Ghana and to Ghanaians in Denmark. Others, like Moses, were adrift, in between, struggling to find their footing and desperately missing home, while still others were there for the exhilarating first time, on a drumming and dancing tour. And then there were those who weren't there, the ones who had not succeeded in leaving, like Jeremiah, Ezekiel, and Gideon, who were still hustling, surviving day by day, and the ones who had died—some of illness at home, a few of sadness abroad. While the last chapter touches upon the transition from early fantasy to subsequent understanding that defines oboroni women's long-term experience of Ghana, here I consider a parallel evolution in Rastas' apprehension of the attainment of the dream of aborokyire, that other beckoning elsewhere.

"The sweet sufferation," Samson had said, when we were both back in Accra in 2009, of the ambivalence of transnationalism. He had spent his first three or four years in England trying to keep a life going in Ghana while getting one going in Europe. Because it was so hard to build a life, you craved the comforts of home, and because you went home, it was harder to build abroad. By 2009, the balance had tilted, but the ambivalence remained. He was in Ghana and hoped to stay for at least six months to finally begin building a house/cultural center in a village outside of Accra, where he had bought land.

But temptations kept coming, calling him back to England: "You go there, you build your life ...." He had a nice group of friends in England, and when in Ghana he missed them. He taught a popular weekly drumming class in a nearby town and had begun teaching drumming in prisons, where he was deeply affected by incarcerated West African men he had met, who had arrived in Europe along routes very different from his. He was traveling to Poland in July as part of a delegation from his town council.

And when he came back to Ghana, his friends here were different. When he had left, the people running his shop thought he "had everything" in England, so they started taking a bit here a bit there, eventually opening their own shops, and now his shop was closed. It was hard to manage the construction project by phone from the United Kingdom. Originally, he had purchased a small plot of land on his own, before teaming up with an American NGO, after which the project grew and then stalled. So now, while they waited for permits and funds to come through, he had decided to build it as he first imagined it: a big house with an office, a workshop, and a performance area. He was thinking of how to live simply but well, even considering farming, so that he could stay longer in Ghana before having to go back to England to earn. But he wanted a Land Rover, so that he could "be free to move" all over Africa.

Looking back upon Rastahood and the mobility it had enabled, the simultaneous losses and gains it engendered, Samson had once told me:

> After some time, when I look back, all the different people that came together—everybody, you know?—we came together for a reason. When somebody is out of it, it's not full, you get me? When we are all together it's full, but when we are not together there is something missing, when I [compare it to] back then. People get missing and come back in, get missing and come back in. So that at different times, different people get missing.

Rastahood had always had a "reason": to be a portal between geographies and existential states, a form betwixt and between locations, a liminality that finally unstuck youth, a transition into adulthood. Rastaworld could not stay "full," not of the same people anyway, for that would mean that it had failed. Enter Rastaworld, and you knew that to succeed, you had to leave it behind.

As I write this in the summer of 2023, it has been 22 years since I first met Simeon, Moses, and Samson in Cape Coast, and 20 years since I began my field-work in Accra. Back then, we were youths in our 20s; now we are parents in middle age. Those Rastas for whom Rastahood had succeeded had transitioned to adult-hood and self-sufficiency, to parenthood and the ability to support their children.

Noah still lives in London. He has three children, all born in England, who have met my son and played with him. He has long renounced his Rastahood. "My little girl always asks me, 'Daddy, how did you come here?' And I tell her everything. One day they will know who their father is. I am trying to establish them, so they don't go through the life that I have gone through." This final sentence was almost word for word one that the carver Godwin Ametewe had said to me of his decision to migrate from his village to Accra a generation earlier.

In a recent interview, Noah reflected on the path that had taken him from Ghana to England, essentially tracing out the path this book has taken. "Greener pastures as we all call it," he said musingly.

> I just wanted a better life. Things have really worked for me. I have my own family, I have children, a wife—I'm pleased with that. I have a really good job—I'm pleased with that. I would say life hasn't disappointed me at all. When I finished secondary school, I didn't have a job, so I couldn't do any-thing but to go into the hustling field, selling drums to tourists. If I had the chance I would have loved to progress to university, but I just couldn't, so that was the only way. I've never regretted it. When you go [to the Arts Cen-tre], that gives you a living at the end of the day. But I wasn't happy with that, because sometimes you make money, and sometimes you don't make money. So I decided to make a move. ... I wanted to explore; I wanted to change my life. I didn't even know that the Arts Centre would lead me to this place. But it just so happened, because I was there—socializing, communicating with tourists—so that opened a door. You know what I mean? Sometimes in life you have to do something that would open a door for you. Because I met so many people, so many Europeans, Americans—all over the world—every-body comes to the Arts Centre, and most of them were my friends. It just so happened that my way for me to come to this place passed through that way.

Like many of the youths who became Rastas, Noah's family had been unable to support his education, precluding him from having a career and middle-class life-style in Ghana. Entering the Arts Centre, hustling and selling drums, had allowed him to survive. But he had wanted more, and then, the people he met there had opened a door, first into an imagined aborokyire, and then into its reality.

This book has attempted to trace this "way" that Rastahood produced, the door that it opened. When Simeon honored his father with a lavish funeral, he was ful-filling the script of adult masculinity, successfully completing the cycle of social reproduction. When Noah saw his daughters in England and nieces in Ghana excelling in school and knew he had spared them from the life he had lived, he was doing the same. So was Samson, when he returned to Ghana to build a home and a cultural center where village children could learn to dance and to use computers. But this same way that had led to these successes had created, in Simeon, a man who almost never came home, in Noah, a man who sent so much money to his extended family in Ghana that he himself had yet to go back, and, in Samson, a man who goes back and forth often, but everywhere and always feels slightly out of place, aware of a distant horizon. Having looked for so long at the beckoning elsewhere of aborokyire, now they looked homeward with longing. Successful migration produces the diasporic condition, and age induces nostalgia. Nonetheless, Rastahood had allowed boys, who in a way had got missing, to come back in as men.

References

Abercrombie, Thomas Alan. 1998. *Pathways of Memory and Power: Ethnography and History among an Andean People*. Madison, WI: University of Wisconsin Press.

Abloh, Frederick. 1967. *Growth of Towns in Ghana: A Study of the Social and Physical Growth of Selected Towns in Ghana*. Ph.D. diss., University of Science and Technology, Kumasi, Ghana.

Abu-Lughod, Lila. 2015. *Do Muslim Women Need Saving?* First Harvard University Press paperback ed. Cambridge, MA: Harvard University Press.

"Academy Of African Music & Arts." n.d. GhanaYello. Accessed May 23, 2023. https://www.ghanayello.com/company/12020/Academy_Of_African_Music_Arts

Achebe, Chinua. 1960. *No Longer at Ease*. London: Heinemann.

Acquah, Ione. 1972. *Accra Survey*. Accra: Ghana Universities Press.

Adamu, Mohammed. 1978. *The Hausa Factor in West African History*. London: Oxford University Press.

Adey, Peter. 2010. *Aerial Life: Spaces, Mobilities, Affects*. Chichester; Malden, MA: Wiley Blackwell.

Adrover, Laura, Daniel Ababio Donkor, and Casey S. McMahon. 2010. "The Ethics and Pragmatics of Making Heritage a Commodity: Ghana's PANAFEST 2009." *TDR: The Drama Review* 54 (2): 155–63.

Agawu, Kofi. 2003. *Representing African Music: Postcolonial Notes, Queries, Positions*. New York and London: Routledge.

Ahmed, Sara. 2010. *The Promise of Happiness*. Durham, NC: Duke University Press.

Akorsu, Angela D. 2013. "Labour Standards Application in the Informal Economy of Ghana: The Patterns and Pressures." *Ekonomski Anali* 196: 157–75. https://doi.org/10.2298/EKA1396157A

Akyeampong, Emmanuel. 2000. "Asante at the Turn of the Twentieth Century." *Ghana Studies* 3 (1): 3–12. https://doi.org/10.3368/gs.3.1.3

Akyeampong, Emmanuel. 2002. "Bukom and the Social History of Boxing in Accra: Warfare and Citizenship in Precolonial Ga Society." *The International Journal of African Historical Studies* 35 (1): 39–60. https://doi.org/10.2307/3097365

Albuquerque, Ana. 1998. "Sex, Beach Boys, and Female Tourists." *Annals of Tourism Research* 25 (4): 948–50.

Alcock, Susan E. 2002. *Archaeologies of the Greek Past: Landscape, Monuments, and Memories*. Cambridge: Cambridge University Press.

Alexander, M. Jacqui, and Chandra Talpade Mohanty. 2012. *Feminist Genealogies, Colonial Legacies, Democratic Futures*. Thinking Gender. New York, NY: Routledge.

Alhassan, Shamara Wyllie. 2020. ""We Stand for Black Livity!": Trodding the Path of Rastafari in Ghana." *Religions* 11 (7): 374.

Alleyne, Mark. 1998. ""Babylon Makes the Rules": The Politics of Reggae Crossover." *Social and Economic Studies* 47 (1): 65–77.

Alleyne, Osei. 2017. *Dancehall Diaspora: Roots, Routes and Reggae Music in Ghana.* Dissertation, ProQuest Dissertations & Theses. University of Pennsylvania.

Allison, Anne. 2013. *Precarious Japan.* Durham, NC: Duke University Press.

Allman, Jean M. 2004. *Fashioning Africa: Power and the Politics of Dress.* Bloomington, IN; Indianapolis, IN: Indiana University Press.

Allman, Jean. 2013. "Kwame Nkrumah, African Studies, and the Politics of Knowledge Production in the Black Star of Africa." *International Journal of African Historical Studies* 46 (2): 181–203.

Amoah, Frank. 1964. *Accra: A Study.* Ph.D. diss., University of Legon, Accra, Ghana.

Annual Report of the Colonies, Gold Coast, 1902. London: His Majesty's Stationary Office.

Annual Report of the Colonies, Gold Coast, 1936–1937. London: His Majesty's Stationary Office.

Antubam, Kofi. 1963. *Ghana's Heritage of Culture.* Leipzig: Koehler & Amelang.

Appadurai, Arjun. 1986. *The Social Life of Things: Commodities in Cultural Perspective.* Cambridge: Cambridge University Press.

Appadurai, Arjun. 1991. "Global Ethnoscapes: Notes and Queries for a Transnational Anthropology Interventions: Anthropologies of the Present." In Richard G. Fox (ed.) *Interventions: Anthropologies of the Present, 191–210.* Santa Fe, NM: School of American Research.

Appiah, Anthony. 2006. *Cosmopolitanism: Ethics in a World of Strangers First ed. Issues of Our Time.* New York: W. W. Norton & Company.

Apter, David E. 1972. *Ghana in Transition.* 2nd revised ed. Princeton Paperbacks. Princeton, NJ: Princeton University Press.

Arom, Simha. 1991. *African Polyphony and Polyrhythm.* Cambridge: Cambridge University Press.

Asempasah, Rogers and Christabel Sam. 2016. Reconstituting the Self: of Names, Discourses and Agency in Amma Darko's Beyond the Horizon. *International Journal of Humanities and Cultural Studies* 2(4):154–168.

Askew, Kelly. 2002. *Performing the Nation: Swahili Music and Cultural Politics in Tanzania,* Vol. 2. Chicago, IL: University of Chicago Press.

Auyero, Javier. 2012. *Patients of the State: The Politics of Waiting in Argentina.* Durham, NC: Duke University Press.

Azoulay, Ariella. 2012. *Civil Imagination: A Political Ontology of Photography.* English language edition. London: Verso.

Azu, Diana Gladys. 1974. *The Ga Family and Social Change.* Vol. 5. African Social Research Documents. Leiden: Afrika-Studiecentrum.

Bakhtin, Mikhail M., Michael Holquist, and Vadim Liapunov. 1990. *Art and Answerability : Early Philosophical Essays.* 1st ed. University of Texas Press Slavic Series, No. 9. Austin: University of Texas Press.

Barber, Karin. 1981. "How Man Makes Gods in West Africa." *Africa* 51 (3): 724–45.

Barber, Karin. 1987. "Literatures in African Languages." *African Affairs* 86 (344): 432–33. https://doi.org/10.1093/oxfordjournals.afraf.a097926

Barber, Karin. 1997. "Preliminary notes on audiences in Africa." *Africa* (pre-2011) 67 (3): 347–362.

Barthes, Roland. 1978. *Image, Music, Text*. 1st ed. Translated by Stephen Heath. New York, NY: Hill & Wang.

Bascom, William. 1976. Changing African Art. In Graburn, Nelson H. H. (ed.) *Ethnic and Tourist Arts: Cultural Expressions from the Fourth World*. 303–319. Berkeley, CA: University of California Press.

Baudrillard, Jean. 1994. *Simulacra and Simulation*. Ann Arbor, MI: University of Michigan Press.

Bauer, Thomas, and Brian McKercher. 2003. *Sex and Tourism: Journeys of Romance, Love, and Lust*. London: The Haworth Hospitality Press.

Bear, Laura. 2014. "Doubt, Conflict, Mediation: The Anthropology of Modern Time." *Journal of the Royal Anthropological Institute* 20: 3–30.

Bear, Laura. 2016. "Time As Technique." *Annual Review of Anthropology* 45: 487–502.

Bebey, Francis. 1975. *African Music: A People's Art. Translated by Josephine Bennett*. 1st U.S. ed. New York, NY: L. Hill.

Beek, Walter E. A. van, and Annette M. Schmidt. 2012. *African Hosts and Their Guests: Cultural Dynamics of Tourism*. Oxford: James Currey.

Beidelman, Thomas O. 1982. *Colonial Evangelism: A Socio-Historical Study of an East African Mission at the Grassroots*. Bloomington, IN: Indiana University Press.

Beidelman, Thomas O. 1997. "Promoting African Art. The Catalogue to the Exhibit of African Art at the Royal Academy of Arts, London." *Anthropos* 92 (1-3): 3–20.

Beidelman, Thomas O. 2012. *The Culture of Colonialism: The Cultural Subjection of Ukaguru*. African Systems of Thought. Bloomington, IN: Indiana University Press.

Ben-Amos, Paula. 1976. "'A La Recherche Du Temps Perdu': On Being an Ebony Carver in Benin." In Graburn, Nelson H. H. (ed.) *Ethnic and Tourist Arts: Cultural Expressions from the Fourth World*. 320–333. Berkeley, CA: University of California Press.

Ben-Amos, Paula. 1977. "Pidgin Languages and Tourist Arts." *Studies in the Anthropology of Visual Communications* 4 (2): 128–39. https://doi.org/10.1525/var.1977.4.2.128

Berkoh, Daniel Kwasi. 1974. *"Urban Primacy in a Developing Country, Ghana: A Case Study of Accra-Tema Metropolitan Area."* Dissertation, Columbia University.

Berlant, Lauren Gail. 2011. *Cruel Optimism*. Durham, NC: Duke University Press.

Berlin, Isaiah. 1971. "Two Concepts of Liberty." In Isaiah Berlin (ed.) *Four Essays on Liberty*. 166–217. New York, NY: Oxford University Press.

Berliner, David. 2005. "An 'Impossible' Transmission: Youth Religious Memories in Guinea Conakry." *American Ethnologist* 32 (4): 576–92.

Bernstein, Elizabeth. 2007. *Temporarily Yours: Intimacy, Authenticity, and the Commerce of Sex*. Chicago, IL: University of Chicago Press.

Berry, Sara S. 1993. *No Condition Is Permanent: The Social Dynamics of Agrarian Change in Sub-Saharan Africa*. Madison, WI: University of Wisconsin Press.

Bird, Charles S., and Martha B. Kendall. 1980. "The Mande Hero: Text and Context." In Ivan Karp and Charles S. Bird (eds.) *Explorations in African Systems of Thought*. Bloomington, IN: Indiana University Press.

Bizas, Eleni. 2014. *Learning Senegalese Sabar: Dancers and Embodiment in New York and Dakar*. 1st ed. New York, NY: Berghahn Books.

Blankson, Samuel. 1973. *Fetu Afahye*. Cape Coast: Printed at the University Press.

Blight, Daniel, and David Roediger. 2019. *The Image of Whiteness: Contemporary Photography and Racialization*. London: SPBH Editions (Self Publish, Be Happy).

Boampong, Owusu. 2015. "Ghanaian Craft Exporters in the Global Market: Binding and Missing Links." *Ghana Journal of Geography* 7 (2): 1–19.

Boorstin, Daniel J. 1964. *The Image: A Guide to Pseudo-Events in America*. New York, NY: Vintage Books, a Division of Random House, Inc.

Bosiwah, Lawrence, Kofi Busia Abrefa, and Charles Okofo Asenso. 2015. "An Etymological Study of the Word 'aborɔfo' (Europeans) and Its Impact on Akan Language." *International Journal of Applied Linguistics and Translation* 1 (1): 1–7.

Botwe-Asamoah, Kwame. 2005. *Kwame Nkrumah's Politico-Cultural Thought and Policies: An African-Centered Paradigm for the Second Phase of the African Revolution*. New York, NY: Routledge.

Bourdieu, Pierre. 1993. *The Field of Cultural Production: Essays on Art and Literature*. European Perspectives. Edited and introduced by Randal Johnson. New York, NY: Columbia University Press.

Bradley, Lloyd. 2001. *This Is Reggae Music: The Story of Jamaica's Music*. New York, NY: Grove Press.

Brand, Richard R. 1972. "The Spatial Organization of Residential Areas in Accra, Ghana, with Particular Reference to Aspects of Modernization." *Economic Geography* 48 (3): 284–98.

Brennan, Denise. 2004. *What's Love Got to Do with It? Transnational Desires and Sex Tourism in the Dominican Republic*. Durham, NC: Duke University Press.

Brown, Jacqueline Nassy. 1998. "Black Liverpool, Black America, and the Gendering of Diasporic Space." *Cultural Anthropology (Wiley-Blackwell)* 13 (3): 291–325.

Brown, Wendy. 2012. "Civilizational delusions: Secularism, tolerance, equality." *Theory & Event* 15 (2): N_A.

Bruner, Edward M. 1996. "Tourism in Ghana: The Representation of Slavery and the Return of the Black Diaspora." *American Anthropologist* 98 (2): 290–304.

Butler, Judith. 1988. "Performative Acts and Gender Constitution: An Essay in Phenomenology and Feminist Theory." *Theatre Journal* 40 (4): 519–31.

Camara, S. 1992. *Gens de la Parole: Essai sur la Condition et le Rôle des Griots dans la Société Malinké*. Paris; Conakry: Karthala and SAEC.

Campt, Tina. 2005. *Other Germans: Black Germans and the Politics of Race, Gender, and Memory in the Third Reich*. 1st pbk. ed. Social History, Popular Culture, and Politics in Germany. Ann Arbor, MI: University of Michigan Press.

Campt, Tina. 2012. *Image Matters: Archive, Photography, and the African Diaspora in Europe*. Durham, NC: Duke University Press.

Carter-Ényì, Quintina, Carter-Ényì, Aaron, and Kevin Nathaniel Hylton. 2019. "How We Got into Drum Circles, and How to Get Out: De-Essentializing African Music." *Intersections: Canadian Journal of Music* 39 (1): 73–92.

Casey, Edward. 1996. "How Do Get from Space to Place in a Fairly Short Stretch of Time: Phenomenological Prolegomena." In Steven Feld and Keith H Basso (eds.) *Senses of Place*. School of American Research Advanced Seminar Series. Santa Fe, NM: School of American Research Press.

Cashmore, Ernest. 1979. *Rastaman: The Rastafarian Movement in England*. Boston, MA: George Allen and Unwin.

Castaldi, Francesca. 2006. *Choreographies of African Identities: Negritude, Dance, and the National Ballet of Senegal*. Champaign, IL: University of Illinois Press.

Charry, Eric. 1996. "A Guide to the Jembe." *Percussive Notes* 34 (2): 66–72.

Charry, Eric. 2000. *Mande Music: Traditional and Modern Music of the Maninka and Mandinka of Western Africa*. Chicago, IL: University of Chicago Press.

Charry, Eric. 2005. "Introduction." In Olatunji, Babatunde (ed.) *The Beat of My Drum: An Autobiography*. Philadelphia, PA: Temple University Press.

Charry, Eric. 2018. "Music and Postcolonial Africa." In Martin S. Shanguhyia and Toyin Falola (eds.) *The Palgrave Handbook of African Colonial and Postcolonial History*, Vol. 2. New York, NY: Palgrave Macmillan.

Chernoff, John Miller. 1979. *African Rhythm and African Sensibility: Aesthetics and Social Action in African Musical Idioms*. A Phoenix Book. Chicago, IL: University of Chicago Press.

Chernoff, John. 1981. *African Rhythm and African Sensibility: Aesthetics and Social Action in African Musical Idioms*. Chicago, IL: University of Chicago Press.

Chernoff, John Miller. 2003. *Hustling Is Not Stealing: Stories of an African Bar Girl*. Chicago, IL: University of Chicago Press.

Chevannes, Barry. 1994. *Rastafari: Roots and Ideology*. 1st ed. Syracuse, NY: Syracuse University Press.

Chevannes, Barry. 1998. *Rastafari and Other African-Caribbean Worldviews*. New Brunswick, NJ: Rutgers University Press.

Christiansen, Christian, Mats Utas, and Henrik Vigh. 2006. *Navigating Youth, Generating Adulthood: Social Becoming in an African Context*. Uppsala: Nordiska Afrikainstitutet.

Chude-Sokei, Louis. 1994. "Post-Nationalist Geographies: Rasta, Ragga, and Reinventing Africa." *African Arts* 27 (4): 80–96.

Clark, Gracia. 1994. *Onions Are My Husband: Survival and Accumulation by West African Market Women*. Chicago, IL: University of Chicago Press.

Clarke, Kamari Maxine, and Deborah A. Thomas, eds. 2006. *Globalization and Race: Transformations in the Cultural Production of Blackness*. Durham, NC: Duke University Press.

Clarke, Sebastian. 1980. *Jah Music: The Evolution of the Popular Jamaican Song*. London: Heinemann Educational Books Ltd.

Clarke, Sebastian. 1994. "Culture, Concept, Aesthetics: The Phenomenon of the African Musical Universe in Western Musical Culture." *African American Review* 29 (2): 329.

Clifford, James. 1997. *Routes: Travel and Translation in the Late Twentieth Century*. Cambridge: Harvard University Press.

Coe, Cati. 2002. "Educating an African Leadership: Achimota and the Teaching of African Culture in the Gold Coast." *Africa Today* 49 (3): 23–44.

Coe, Cati. 2005. *Dilemmas of Culture in African Schools: Youth, Nationalism, and the Transformation of Knowledge*. Chicago, IL: University of Chicago Press.

Coe, Cati. 2011. *Everyday Ruptures: Children, Youth, and Migration in Global Perspective*. Nashville, TN: Vanderbilt University Press.

Coe, Cati. 2014. "The Divergence of Art from 'African Culture': From Achimota to the Educational Reforms of 1986 in Ghana's Art Education." *Critical Interventions* 8 (1): 52–73.

Cohen, Abner. 1969. *Custom & Politics in Urban Africa: A Study of Hausa Migrants in Yoruba Towns*. Berkeley, CA: University of California Press.

Cohen, Joel. 2012. "Stages in Transition: Les Ballets Africains and Independence, 1959 to 1960." *Journal of Black Studies* 43 (1): 11–48.

Cole, Jennifer. 2010. *Sex and Salvation: Imagining the Future in Madagascar*. Chicago, IL: University of Chicago Press.

Cole, Jennifer, and Deborah L. Durham. 2007. *Generations and Globalization: Youth, Age, and Family in the New World Economy*. Bloomington, IN: Indiana University Press.

Cole, Jennifer, and Lynn M Thomas. 2009. *Love in Africa*. Chicago, IL: University of Chicago Press.

Colen, Shellee. 1995. "'Like a Mother to Them': Stratified Reproduction and West Indian Childcare Workers and Employers in New York." In Faye D. Ginsburg and Rayna Rapp (eds.) *Conceiving the New World Order: The Global Politics of Reproduction*. Berkeley, CA: University of California Press.

Collins, John. 2001. "Making Ghanaian Music Exportable." Paper presented at the Ghanaian Music Awards, Accra.

Comaroff, John L, and Jean Comaroff. 1992. *Ethnography and the Historical Imagination.* Studies in the Ethnographic Imagination. Boulder, CO: Westview Press.

Comaroff, Jean, and John L. Comaroff. 2001. *Millennial Capitalism and the Culture of Neoliberalism.* Durham, NC: Duke University Press.

Comaroff, Jean, and John L. Comaroff. 2012. "Theory from the South: Or, How Euro-America Is Evolving Toward Africa." *Anthropological Forum* 22 (2): 113–31.

Constable, Nicole. 2009. "The Commodification of Intimacy: Marriage, Sex, and Reproductive Labor." *Annual Review of Anthropology* 38 (1): 49–64.

Constable, Nicole. 2016. "Reproductive Labor at the Intersection of Three Intimate Industries: Domestic Work, Sex Tourism, and Adoption." *Positions* 24 (1): 45–69.

Cooper, Frederick. 2002. *Africa since 1940: The Past of the Present.* New Approaches to African History. Cambridge: Cambridge University Press.

Cutter, Charles H. 1971. *Nation-Building in Mali: Art, Radio, and Leadership in a Pre-Literate Society.* Los Angeles, CA: University of California.

Dahles, Heidi, and Karin Bras. 1999. "Entrepreneurs in Romance: Tourism in Indonesia." *Annals of Tourism Research* 26 (2): 267–93.

Dakubu, M. E. Kropp. 1972. "Linguistic Pre-History and Historical Reconstruction: The Ga-Adangme Migrations." *Transactions of the Historical Society of Ghana* 13 (1): 87–111.

Dakubu, Mary Esther Kropp. 1997. *Korle Meets the Sea: A Sociolinguistic History of Accra.* New York, NY: Oxford University Press.

Davis, Stephen. 1982. *Reggae International.* New York, NY: Rogner and Bernhard.

de Albuquerque, Klaus. 1998a. "In Search of the Big Bamboo." *Transition* (77): 48–57.

de Albuquerque, Klaus. 1998b. "Sex, Beach Boys, and Female Tourists in the Caribbean." *Sexuality and Culture* 2: 87–111.

Dei, George. 2012. "Reclaiming Our Africanness in the Diasporized Context: The Challenge of Asserting a Critical African Personality." *The Journal of Pan African Studies* 4 (10): 42–57.

Deveaux, Monique. 2018. "Appeals to Choice and Sexual Equality." In Marie Claire Foblets, Michele Graziadei, and Alison Dundes Renteln (eds.) *Personal Autonomy in Plural Societies: A Principle and Its Paradoxes.* Law and Anthropology Series. Abingdon: Oxon: Routledge, an imprint of the Taylor & Francis Group.

Dhillon, Navtej, and Tarik Yousef. 2009. *Generation in Waiting: The Unfulfilled Promise of Young People in the Middle East.* Washington, DC: Brookings Institution Press.

Diouf, Sylviane A. 2003. *Fighting the Slave Trade: West African Strategies* Western African Studies. Athens, OH: Ohio University Press.

Donkor, David A. 2016. *Spiders of the Market: Ghanaian Trickster Performance in a Web of Neoliberalism.* Bloomington, IN; Indianapolis, IN: Indiana University Press.

Dor, George. 2004. "Communal Creativity and Song Ownership in Anlo Ewe Musical Practice: The Case of Havolu." *Ethnomusicology* 48 (1): 26–51.

Dor, George Worlasi Kwasi. 2014. *West African Drumming and Dance in North American Universities: An Ethnomusicological Perspective.* Jackson, MS: University Press of Mississippi.

Dovlo, Elom. 2002. "Rastafari, African Hebrews & Black Muslims: Return 'Home' Movements in Ghana." *Exchange* 31 (1): 2–22.

Durham, Deborah. 2000. "Youth and the Social Imagination in Africa: Introduction to Parts 1 and 2." *Anthropological Quarterly* 73 (3): 113–20. https://doi.org/10.1353/anq.2000.0003

Dyer, Richard. 1997. *White: Essays on Race and Culture.* London: Routledge.

Ebere, Charles, and Evan Charles-Ebere. 2018. "Bumster Subculture and Rastafari Identity in the Gambia: The Search for Survival and Social Mobility among Marginalised Male Youth." *Bangladesh e-Journal of Sociology* 15 (1): 134–47.

Ebron, Paula. 1997. "Traffic in Men." In Maria Luise Grosz-Ngaté and Omari H Kokole (eds.) *Gendered Encounters: Challenging Cultural Boundaries and Social Hierarchies in Africa*. New York, NY: Routledge.

Ebron, Paula. 2002. *Performing Africa*. Princeton, NJ: Princeton University Press.

Edelman, Lee. 2004. *No Future: Queer Theory and the Death Drive*. Durham, NC: Duke University Press.

Edmonds, E. B. 1998. "The Structure and Ethos of Rastafari." In N. S. Murrell, W. D. Spencer, and A. A. McFarlane (eds.) *Chanting Down Babylon: The Rastafari Reader*. Philadelphia, PA: Temple University Press.

Edmonds, Ennis B. 2003. *Rastafari: From Outcasts to Culture Bearers*. New York, NY: Oxford University Press.

Edmondson, Laura. 2007. *Performance and Politics in Tanzania: The Nation on Stage*. African Expressive Cultures. Bloomington, IN: Indiana University Press.

Emery, Lynne Fauley. 1988. *Black Dance: From 1619 to Today*, 2nd rev. ed. Princeton, NJ: Princeton University Press.

England, Paula. 2005. "Emerging Theories of Care Work." *Annual Review of Sociology* 31: 381–99.

Epprecht, Marc, and Vasu Sigamoney. 2013. "Meanings of Homosexuality, Same-Sex Sexuality, and Africanness in Two South African Townships: An Evidence-Based Approach for Rethinking Same-Sex Prejudice." *African Studies Review* 56 (2): 83–107.

Erlmann, Veit. 1996. *Nightsong: Performance, Power, and Practice in South Africa*. Chicago, IL: University of Chicago Press.

Fabian, Johannes. 1998. "Curios and Curiosity: Notes on Reading Torday and Frobenius." Cambridge: Cambridge University Press.

Faier, Lieba. 2007. "Filipina Migrants in Rural Japan and Their Professions of Love." *American Ethnologist* 34: 148–162.

Fanon, Frantz. 1967. *Black Skin, White Masks*. New York, NY: Grove Press.

Fanon, Frantz. 1965. *The Wretched of the Earth*. New York, NY: Grove Press, Inc.

Federici, Silvia. 2020. *Revolution at Point Zero: Housework, Reproduction, and Feminist Struggle* (version Second edition.). 2nd ed. Oakland, CA: PM Press.

Feld, Steven. 1988. "Aesthetics as Iconicity of Style, or 'Lift-Up-Over Sounding': Getting into the Kaluli Groove." *Yearbook for Traditional Music* 20: 74–113.

Feld, Steven. 1995. "Chapter 3: From Schizophonia to Schismogenesis: The Discourses and Practices of World Music and World Beat." In George E. Marcus and Fred R. Myers., *Traffic in Culture: Refiguring Art and Anthropology*. 96–126. Berkeley, CA: University of California Press.

Feld, Steven. 2000. "A Sweet Lullaby for World Music." *Public Culture* 12 (1): 145–71.

Feld, Steven. 2012. *Jazz Cosmopolitanism in Accra: Five Musical Years in Ghana*. Durham, NC: Duke University Press.

Ferguson, James. 1999. *Expectations of Modernity: Myths and Meanings of Urban Life on the Zambian Copperbelt*. Berkeley, CA: University of California Press.

Ferguson, James. 2006. *Global Shadows: Africa in the Neoliberal World Order*. Durham, NC: Duke University Press.

Field, Margaret Joyce. 1937. *Religion and Medicine of the Gā People*. Accra: Presbyterian Book Depot.

Field, Margaret Joyce. 1940. *Social Organization of the Ga People*. London: Crown Agents for the Colonies.

Flaig, Valérie Hartwich. 2010. *The Politics of Representation and Transmission in the Globalization of Guinea's Djembe*. Ph.D. diss., University of Michigan, Ann Arbor, MI.

Folbre, Nancy, ed. 2012. *For Love or Money: Care Provision in the United States*. New York, NY: Russell Sage Foundation.

Förster, Till. 2013. "Work and Workshop. The iteration of style and genre in two workshop settings. Côte d'Ivoire and Cameroon." In Kasfir, Sidney Littlefield, and Förster Till, eds. *African Art and Agency in the Workshop (Bloomington 2013)*.

Foster, Philip. 1965. "The Vocational School Fallacy in Development Planning." In A. Anderson and M. Bowman (eds.) *Education and Economic Development*. 142–166. Chicago, IL: Aldine.

Fosu, Kwaku Amoako-Attah. 1999. *Festivals in Ghana*. Kumasi: KAA Fosu.

Frankenberg, Ruth. 1993. *White Women, Race Matters: The Social Construction of Whiteness*. Minneapolis, MN: University of Minnesota Press.

Frankenburg, Ruth, ed. 1997. *Displacing Whiteness: Essays in Social and Cultural Criticism*. Durham, NC: Duke University Press.

Freeman, Carla. 2000. *High Tech and High Heels in the Global Economy: Women, Work, and Pink-Collar Identities in the Caribbean*. Durham, NC: Duke University Press.

Freeman, Carla. 2007. "The 'Reputation' of Neoliberalism." *American Ethnologist* 34 (2): 252–67.

Frohlick, Susan. 2007. "Fluid Exchanges: The Negotiation of Intimacy between Tourist Women and Local Men in a Transnational Town in Caribbean Costa Rica." *CISO City & Society* 19 (1): 139–68.

Frohlick, Susan. 2010. "Sex, Beach Boys, and Female Tourists in the Caribbean." *Sexuality and Culture* 2: 87–111.

Frohlick, Susan. 2013. *Sexuality, Women, and Tourism: Cross-Border Desires through Contemporary Travel*. London; New York, NY: Routledge.

Fuller, Harcourt. 2014. *Building the Ghanaian Nation-State: Kwame Nkrumah's Symbolic Nationalism*. New York, NY: Palgrave Macmillan.

Fyfe, Christopher. 1992. "Race, Empire and the Historians." *Race & Class* 33 (4): 15–30.

Gable, Eric. 2000. "The Culture Development Club: Youth, Neo-Tradition, and the Construction of Society in Guinea-Bissau." *Anthropological Quarterly* 73 (4): 195–203.

Gaines, Kevin Kelly. 2006. *American Africans in Ghana: Black Expatriates and the Civil Rights Era*. The John Hope Franklin Series in African American History and Culture. Chapel Hill, NC: University of North Carolina Press.

Gates, Henry Louis Jr. 1989. *The Signifying Monkey: A Theory of African-American Literary Criticism*. New York, NY: Oxford University Press.

Gaudette, Pascal. 2013. "*Jembe* Hero: West African Drummers, Global Mobility and Cosmopolitanism as Status." *Journal of Ethnic and Migration Studies* 39 (2): 295–310.

Geertz, Clifford. 1983. "Art as a Cultural System." In *Local Knowledge: Further Essays in Interpretive Anthropology*, 94–120. New York, NY: Basic Books.

Gell, Alfred. 1998. *Art and Agency: An Anthropological Theory*. Oxford: Clarendon Press.

Gell, Alfred. 2001. *The Anthropology of Time: Cultural Constructions of Temporal Maps and Images*. Explorations in Anthropology. Oxford: Berg.

"'Ghana's Bob Marley' spreads message of brotherhood." African Voices, CNN August 23, 2011. www.cnn.com. Accessed March 13, 2021.

"Ghana Forestry Commission – Ghana Wood Products, Timber Products - Species." n.d. Accessed June 21, 2023. https://www.ghanatimber.org/species.php.

Gilroy, Paul. 1987. *"There Ain't No Black in the Union Jack": The Cultural Politics of Race and Nation*. Black Literature and Culture. Chicago, IL: University of Chicago Press.

Gilroy, Paul. 1993. *The Black Atlantic: Modernity and Double Consciousness*. Cambridge, MA: Harvard University Press.

Gilvin, Amy. 2014. "Converging Pedagogies in African Art Education Colonial Legacies and Post-Independence Aspirations." *Critical Interventions* 8 (1): 1–9.

Ginsburg, Faye. 1995. "The Parallax Effect: The Impact of Aboriginal Media on Ethnographic Film." *Visual Anthropology Review* 11 (2): 64–76.

Ginsburg, Faye, and Rayna Rapp. 1991. "The Politics of Reproduction." *Annual Review of Anthropology* 20 (1): 311–43. https://doi.org/10.1146/annurev.an.20.100191.001523

Glenn, Evelyn Nakano. 2012. *Forced to Care: Coercion and Caregiving in America*. Cambridge, MA: Harvard University Press.

Glissant, Edouard. 2002. *The Cry of the World: Treatise on the Whole-World*. Liverpool: Liverpool University Press.

Gocking, Roger. 2000. "The Tribunal System in Ghana's Fourth Republic: An Experiment in Judicial Reintegration." *African Affairs* 99 (394): 47–71.

Goffman, Erving. 1959. *The Presentation of Self in Everyday Life*. New York, NY: Anchor.

Goffman, Erving. 1974. *Frame Analysis*. New York, NY: Harper & Row.

Gordimer, Nadine. 2007. "Beethoven Was One-Sixteenth Black." In *Beethoven Was One-Sixteenth Black And Other Stories*. 1st ed. New York, NY: Farrar, Straus and Giroux.

Gott, Elisha S., and Katherine Loughran. 2010. *Contemporary African Fashion*. Bloomington, IN; Indianapolis, IN: Indiana University Press.

Graburn, Nelson H. H. 1976. *Ethnic and Tourist Arts: Cultural Expressions from the Fourth World*. Berkeley, CA: University of California Press.

Graburn, Nelson H. H. 1989. "Tourism: the Sacred Journey." In Valene L. Smith (ed.) *Hosts and Guests, the Anthropology of Tourism*. Philadelphia, PA: University of Pennsylvania Press.

Grant, Richard. 1999. "Economic Globalisation: Politics and Trade Policy in Ghana and Kenya." *Geopolitics* 4 (1): 57–82.

Grossberg, Lawrence, Stuart Hall, and Paul Du Gay. 1996. "Identity and Cultural Studies: Is That All There Is." In Du Gay, Paul, and Stuart Hall. 2011, *Questions of Cultural Identity*. SAGE.

Guilbault, Jocelyne. 1993. "On Redefining the 'Local' through World Music." *The World of Music* 35 (2): 33–47.

Guyer, Jane I. 2007. "Prophecy and the Near Future: Thoughts on Macroeconomic, Evangelical, and Punctuated Time." *American Ethnologist* 34 (3): 409–21.

Hale, Thomas A. 1998. *Griots and Griottes: Masters of Words and Music*. Bloomington, IN: Indiana University Press.

Hall, Stuart. 1990. "Cultural Identity and Diaspora." In Jonathon Rutherford (ed) *Identity, Community, Culture, Difference*, 222–37. London: Lawrence and Wishart.

Hall, Stuart. 1993. "What Is This Black in Black Popular Culture." *Social Justice* 20 (1/2): 104–14.

Hannerz, Ulf. 1992. *Cultural Complexity: Studies in the Social Organization of Meaning*. New York, NY: Columbia University Press.

Harding, Frances. 2013. *The Performance Arts in Africa: A Reader*. London: Routledge.

Hardt, Michael. 1999. "Affective Labor." *Boundary* 26 (2): 89–99.

Hardt, Michael, and Antonio Negri. 2000. "Introduction." *Discourse* 22 (3): 3. https://doi.org/10.1353/dis.2000.0005

Harney, Elizabeth. 2004. *In Senghor's Shadow: Art, Politics, and the Avant-Garde in Senegal, 1960–1995*. Durham, NC: Duke University Press.

Harris, Cheryl. 1993. "Whiteness as Property." *Harvard Law Review* 106 (8): 1710–69.

Harrod, Tanya. 1989. "The 'Breath of Reality': Michael Cardew and the Development of Studio Pottery in the 1930s and 1940s." *Journal of Design History* 2 (2–3): 145–59.

Hart, Keith. 1973. "Informal Income Opportunities and Urban Employment in Ghana." *The Journal of Modern African Studies* 11 (1): 61–89.

Hartman, Saidiya V. 2008. *Lose Your Mother: A Journey Along the Atlantic Slave Route*. New York, NY: Farrar, Straus & Giroux.

Harvey, David. 1990. *The Condition of Postmodernity: An Enquiry into the Origins of Cultural Change*. Oxford England: Blackwell.

Hebdige, Dick. 1979. *Subculture: The Meaning of Style*. London: Methuen.

Hendrickson, Hildi. 1996. "Introduction." In Hildi Hendrickson, ed. *Clothing and Difference: Embodied Identities in Colonial and Post-Colonial Africa*. Durham, NC: Duke University Press.

Herppich, Birgit. 2016. *Pitfalls of Trained Incapacity: The Unintended Effects of Integral Missionary Training in the Basel Mission on Its Early Work in Ghana (1828–1840)*. American Society of Missiology Monograph Series, Vol. 26. Cambridge: James Clarke.

Herz-Jakoby, Anja. 2013. "Mobility in the Life and Work of Contemporary Ghanaian Artists." *Critical Interventions* 7 (1): 65–78.

Hess, Janet Berry. 2000. "Imagining Architecture: The Structure of Nationalism in Accra, Ghana." *Africa Today* 47 (2): 35–58.

Hess, Janet Berry. 2003. "Imagining Architecture II: 'Treasure Storehouses' and Constructions of Asante Regional Hegemony." *Africa Today* 50 (1): 26–48.

Himpele, Jeff. 2002. "Arrival Scenes: Complicity and Media Ethnography in the Bolivian Public Sphere." In Faye Ginsburg, Lila Abu-Lughod, and Brian Larkin (eds.) *Media Worlds: Anthropology on New Terrain*. Berkeley, CA: University of California Press.

Hirschmann, Nancy J. 1996. "Toward a Feminist Theory of Freedom." *Political Theory* 24 (1): 46–67.

Hobsbaum, Eric, and Terence Ranger, eds. 1983. *The Invention of Tradition*. Cambridge; New York, NY: Cambridge University Press.

Hochschild, Arlie. 1983. *The Managed Heart: Commercialization of Human Feeling*. Berkeley, CA: University of California Press.

Holsey, Bayo. 2008. *Routes of Remembrance: Refashioning the Slave Trade in Ghana*. Chicago, IL: University of Chicago Press.

Hondagneu-Sotelo, Pierrette. 2001. *Doméstica: Immigrant Workers Cleaning and Caring in the Shadows of Affluence*. Berkeley, CA: University of California Press.

Hondagneu-Sotelo, Pierrette, and Ernestine Avila. 1997. "'I'm Here, but I'm There': The Meanings of Latina Transnational Motherhood." *Gender and Society* 11 (5): 548–71.

Honwana, Alcinda M. 2012. *The Time of Youth: Work, Social Change, and Politics in Africa*. Sterling, VA: Kumarian Press.

Honwana, Alcinda M., and Filip de Boeck. 2005. *Makers & Breakers: Children & Youth in Postcolonial Africa*. Oxford; Trenton, NJ; Dakar: James Currey; Africa World Press; Codesria.

hooks, bell. 1992. *Black Looks: Race and Representation*. Boston, MA: South End Press.

Hopkins, Nicholas S. 1965. "Le Théâtre Moderne au Mali." *Présence Africaine* 53 (1): 162–93.

Hull, Arthur. 1998. *Drum Circle Spirit: Facilitating Human Potential through Rhythm*. Performance in World Music Series, No. 12. Gilsum, NH: White Cliffs Media.

Hyden, G. 2006. *African Politics in Comparative Perspective*. Cambridge, MA: Cambridge University Press.

Information Hood. 2022. "Academy Of African Music & Arts Contact Details." *Information Hood* (blog). December 7, 2022. https://www.informationhood.com/academy-of-african music-arts-contact-details/

Jackson, John L. 2001. *Harlemworld: Doing Race and Class in Contemporary Black America*. Chicago, IL; London: University of Chicago Press.

Jackson, John L. 2005. *Real Black: Adventures in Racial Sincerity*. Chicago, IL: University of Chicago Press.

Jackson, Ian, and Robert Assasie Oppong. 2014. "The Planning of Late Colonial Village Housing in the Tropics: Tema Manhean, Ghana." *Planning Perspectives* 29 (4): 475–99.

Jackson, Robert H, and Carl G Rosberg. 1984. "Personal Rule: Theory and Practice in Africa." *Comparative Politics* 16 (4): 421–42.

Jacobs, Jessica. 2009. "Have Sex Will Travel: Romantic 'Sex Tourism' and Women Negotiating Modernity in the Sinai." *Gender, Place & Culture* 16 (1): 43–61.

Jameson, Fredric. 1991. *Postmodernism, or, the Cultural Logic of Late Capitalism*. Post-Contemporary Interventions. Durham, NC: Duke University Press.

Järvenpää, Tuulikki. 2016. "From Gugulethu to the World: Rastafarian Cosmopolitanism in the South African Reggae Music of Teba Shumba and the Champions." *Popular Music and Society* 40 (4): 1–21.

Jeffrey, Craig. 2010. "Timepass: Youth, Class, and Time among Unemployed Young Men in India." *American Ethnologist* 37 (3): 465–81.

Jeffreys, Sheila. 2003. "Sex Tourism: Do Women Do It Too?" *Leisure Studies* 22 (3): 223–38.

Jenkins, David. 1975. *Black Zion: The Return of Afro-Americans and West Indians to Africa*. London: Wildwood House.

Jenkins, Ray. 1994. "William Ofori Atta, Nnambi Azikiwe, J.B. Danquah and the "Grilling" of W.E.F. Ward of Achimota in 1935." *History in Africa* 21:171–189.

Johnson, Marion. 1979. "Ashanti Craft Organization." *African Arts* 13 (1): 60–63.

Johnson-Hanks, Jennifer. 2006. *Uncertain Honor: Modern Motherhood in an African Crisis*. Chicago, IL: University of Chicago Press.

Johnson-Hanks, Jennifer. 2007. "Women on the Market: Marriage, Consumption, and the Internet in Urban Cameroon." *American Ethnologist* 34 (4): 642–58.

Jules-Rosette, Bennetta. 1984. *The Messages of Tourist Art: An African Semiotic System in Comparative Perspective*. Topics in Contemporary Semiotics. New York, NY: Plenum Press.

Kalani. 2002. *All About Jembe Everything You Need to Start Playing Now!* Alfred's World Percussion Series. Van Nuys, CA: Alfred Pub.

Kalani. 2004. *Together in Rhythm: A Facilitator's Guide to Drum Circle Music*. Los Angeles, CA: Alfred Pub. Co.

Kasfir, Sidney L. 2004. "Tourist Aesthetics in the Global Flow: Orientalism and 'Warrior Theatre'" on the Swahili Coast." *Visual Anthropology* 17 (3–4): 319–43.

Kasfir, Sidney Littlefield. 2007. *African Art and the Colonial Encounter: Inventing a Global Commodity*. Bloomington, IN; Indianapolis, IN: Indiana University Press.

Kasfir, Sidney Littlefield, and Till Förster. 2013. *African Art and Agency in the Workshop*. Bloomington, IN; Indianapolis, IN: Indiana University Press.

Keane, Webb. 2003. "Semiotics and the Social Analysis of Material Things." *Language & Communication* 23 (3–4): 409–25.

Kelsky, Karen. 2001. *Women on the Verge: Japanese Women, Western Dreams*. Durham, NC: Duke University Press.

Kempadoo, Kamala. 1999. *Sun, Sex, and Gold: Tourism and Sex Work in the Caribbean.* Lanham, MD: Rowman & Littlefield Publishers.

Kibicho, Wanjohi. 2016. *Sex Tourism in Africa: Kenya's Booming Industry.* New Directions in Tourism Analysis. London: Routledge.

Kilson, Marion. 1974. *African Urban Kinsme : The Ga of Central Accra.* London: C. Hurst.

King, Kenneth. 1969. "Africa and the Southern States of the U.S.A.: Notes on J.H. Oldham and American Negro Education for Africans." *The Journal of African History* 10 (4): 659–77.

King, Kenneth J. 1970. "African Students in Negro American Colleges: Notes on the Good African." *Phylon (1960-)* 31 (1): 16–30.

King, Kenneth. 1971. *Western Education and Political Domination in Africa: A Study in Critical and Dialogical Pedagogy.* London: Methuen.

King, Stephen A. 1999. "The Co-Optation of a 'Revolution': Rastafari, Reggae, and the Rhetoric of Social Control." *Howard Journal of Communication* 10 (2): 77–95.

King, Stephen A., Brian T. Bays, and Philip R. Foster. 2002. *Reggae, Rastafari, and the Rhetoric of Social Control.* Jackson, MS: University Press of Mississippi.

Kingsley, Mary H. 1971 [1897]. *Travels in West Africa: Congo Français, Corisco and Cameroons.* New York, NY: Macmillan.

Knight, Roderic. 1984. "Music in Africa: The Manding Contexts." In Gerard Behague (ed.) *Performance Practice: Ethnomusicological Perspectives*, 53–90. Westport, CT: Greenwood Press.

Konadu-Agyemang, Kwabena. 2001. *The Political Economy of Housing and Urban Development in Africa: Ghana's Experience from Colonial Times to 1998.* Westport, CT: Praeger.

Kwakye-Opong, Regina. 2014. "Clothing and colour symbolisms in the Homowo festival: A means to sociocultural development." *Research on Humanities and Social Sciences* 4 (13): 112–25.

Kwami, Atta. 2013. "Kofi Antubam, 1922–1964: A Modern Ghanaian Artist, Educator, and Writer." *A Companion to Modern African Art.* Oxford: John Wiley & Sons.

Lake, Obiagele. 1995. "Toward a Pan-African Identity: Diaspora African Repatriates in Ghana." *Anthropological Quarterly* 68 (1): 21–36.

Lave, Jean, and Etienne Wenger. 1991. *Situated Learning: Legitimate Peripheral Participation.* Cambridge: Cambridge University Press.

Lawrance, Benjamin Nicholas. 2007. *Locality, Mobility, and "Nation": Periurban Colonialism in Togo's Eweland, 1900–1960.* Vol. 31. Rochester, NY: University Rochester Press.

Leach, Edmund R. 1958. "Magical Hair." *The Journal of the Royal Anthropological Institute of Great Britain and Ireland* 88 (2): 147–64.

Lentz, Carola. 2001. "Local Culture in the National Arena: The Politics of Cultural Festivals in Ghana." *African Studies Review* 44 (3): 47–72.

Lindstrom, Lamont. 1981. "'Big Man:' a Short Terminological History." *American Anthropologist* 83 (4): 900–905.

Lucht, Hans. 2011. *Darkness before Daybreak.* Berkeley, CA: University of California Press.

Lugard, Frederick D. 1923. *The Dual Mandate in British Tropical Africa.* 5th ed. London: Frank Cass & Co. Ltd.

Lugard, Frederick D. 1965. *The Dual Mandate in British Tropical Africa.* London: F. Cass.

Lynch, Kevin. 1960. *The Image of the City.* Cambridge, MA: MIT Press.

Mabogunje, Akin L. 1990. "Urban Planning and the Post-Colonial State in Africa: A Research Overview." *African Studies Review* 33 (2): 121–203.

MacCannell, Dean. 1976. *The Tourist: A New Theory of the Leisure Class*. New York, NY: Schocken Books.

Mafeje, Archie. 2000. "Africanity: A Combative Ontology." CODESRIA Bulletin 1 & 4: b66–71.

Mahoney, Dillon. 2017. *The Art of Connection: Risk, Mobility, and the Crafting of Transparency in Coastal Kenya*. Oakland, CA: University of California Press.

Mains, Daniel. 2007. "Neoliberal Times: Progress, Boredom, and Shame among Young Men in Urban Ethiopia." *American Ethnologist* 34 (4): 659–73.

Mamdani, Mahmood. 1996. "Indirect Rule, Civil Society, and Ethnicity: The African Dilemma." *Social Justice -San Francisco* 23 (1/2): 145–50.

Mamdani, Mahmood. 1999. "Historicizing Power and Responses to Power: Indirect Rule and Its Reform." *Social Research* 66 (3): 859–86.

Marcus, George E., and Fred R. Myers. 1995. *The Traffic in Culture: Refiguring Art and Anthropology*. Berkeley, CA: University of California Press.

Martin, Andrew, and George Ross. 1999. *The Brave New World of European Labor: European Trade Unions at the Millennium*. New York, NY: Berghahn Books.

Masquelier, Adeline. 2013. "Teatime: Boredom and the Temporalities of Young Men in Niger." *Africa: The Journal of the International African Institute* 83 (3): 385–402.

Massey, Doreen. 2005. *For Space*. London: Sage.

Mauss, Marcel. 1967. *The Gift: Forms and Functions of Exchange in Archaic Societies*. New York, NY: Norton.

Mbembe, Achille. 2001. *On the Postcolony*. Berkeley, CA: University of California Press.

McClintock, Anne. 1995. *Imperial Leather: Race, Gender, and Sexuality in the Colonial Contest*. New York, NY: Routledge.

McCombes, Lucy. 2012. "Host-Guest Encounters in a Gambian 'Love' Bubble." In Beek, Walter E. A. van, and Annette M. Schmidt (eds.) *African Hosts and Their Guests: Cultural Dynamics of Tourism*. Oxford: James Currey.

McLeod, Malcolm D. 1981. *The Asante*. London: British Museum Publications.

McWhorter, Ladelle. 2013. "Post-Liberation Feminism and Practices of Freedom." *Foucault Studies* 16: 54–73.

Meisch, Lynn A. 1995. "Gringas and Otavaleños: Changing Tourist Relations." *Annals of Tourism Research* 22 (2): 441–62.

Meiu, George Paul. 2009. "'Mombasa Morans': Embodiment, Sexual Morality, and Samburu Men in Kenya." *Canadian Journal of African Studies/Revue Canadienne Des Études Africaines* 43 (1): 105–28.

Meiu, George. 2017. *Ethno-Erotic Economies: Sexuality, Money, and Belonging in Kenya*. Chicago, IL: University of Chicago Press.

Mercer, Kobena. 1987. "Black Hair/Style Politics." *New Formations* 3 (Winter): 33–54.

Mercer, Kobena. 2016. "Diaspora Aesthetics and Visual Culture." In *Travel & See: Black Diaspora Art Practices since the 1980s*, 227–247. New York, NY: Duke University Press.

Meyer, Birgit. 1999. *Translating the Devil: Religion and Modernity among the Ewe in Ghana*. International African Library, 21. Edinburgh: Edinburgh University Press for the International African Institute.

Meyer, Birgit. 1998. "'Make a Complete Break with the Past.' Memory and Post-Colonial Modernity in Ghanaian Pentecostalist Discourse." *Journal of Religion in Africa* 28 (3): 316–49.

Meyer, Birgit. 2004. "'Praise the Lord': Popular Cinema and Pentecostalite Style in Ghana's New Public Sphere." *American Ethnologist* 31 (1): 92–110.

Middleton, Darren. 2006. "As It Is in Zion: Seeking the Rastafari in Ghana, West Africa." *Black Theology* 4 (2): 151–72.

Millar, Kathleen M. 2014. "The Precarious Present: Wageless Labor and Disrupted Life in Rio de Janeiro, Brazil." *Cultural Anthropology* 29 (1): 32–53.

Miller, Monica L. 2009. *Slaves to Fashion: Black Dandyism and the Styling of Black Diasporic Identity*. Durham, NC: Duke University Press.

Milou, Niëns. 2010. "'Before Whites Were Devils, Now We Are Marrying Them': Moral and Racial Discourses on Romance Tourism in the Gambia." *Mande Studies* 12: 121–142.

Molé, Noelle J. 2010. "Precarious Subjects: Anticipating Neoliberalism in Northern Italy's Workplace." *American Anthropologist* 112 (1): 38–53.

Mudimbe, Valentin-Yves. 1988. *The Invention of Africa: Gnosis, Philosophy, and the Order of Knowledge*. Bloomington, IN; Indianapolis, IN: Indiana University Press.

Mudimbe, Valentin-Yves. 1992. *The Surreptitious Speech: Présence Africaine and the Politics of Otherness, 1947–1987*. Chicago, IL: University of Chicago Press.

Mullings, Beverley. 2000. "Fantasy tours: exploring the global consumption of Caribbean sex tourisms." In Gottdiener, Mark (eds.) *New Forms of Consumption: Consumers, Culture, and Commodification*, 227–250. Lanham, MD: Rowman & Littlefield.

Munn, Nancy D. 1992. "The Cultural Anthropology of Time: A Critical Essay." *Annual Review of Anthropology* 21: 93–123.

Murrell, Nathaniel Samuel, William David Spencer, and Adrian Anthony McFarlane (eds.) 1998. *Chanting Down Babylon: The Rastafari Reader*. Philadelphia, PA: Temple University Press.

Myers, Fred. 1995. "Representing Culture: the Production of Discourse(s) for Aboriginal Acrylic Paintings." In George Marcus and Fred Myers (eds.) *The Traffic in Culture*, 55–95. Berkeley, CA: University of California Press.

Myers, Fred. 2002. *Painting Culture: The Making of an Aboriginal High Art*. Durham, NC: Duke University Press.

Nagel, Joane. 2000. "States of Arousal/Fantasy Islands: Race, Sex, and Romance in the Global Economy of Desire." *American Studies* 41 (2–3): 159–81.

Nagel, Joane. 2003. *Race, Ethnicity, and Sexuality: Intimate Intersections, Forbidden Frontiers*. New York, NY: Oxford University Press.

"Nani Agbeli – Professional in Ghanaian, West African Music and Dance." n.d. Drum Dancewith Nani. Accessed June 21, 2023. https://www.naniagbeli.com

Negri, Antonio. 2003. *Time for Revolution*. London: Continuum.

Neilson, Brett, and Ned Rossiter. 2008. "Precarity as a Political Concept, or Fordism as Exception." *Theory, Culture & Society* 25 (7–8): 51–72.

Nketia, J. H. Kwabena. 1963. *African Music in Ghana*. Northwestern University African Studies, No. 11. Evanston, IL: Northwestern University Press.

Nketia, J. H. Kwabena. 1974. *The Music of Africa*. New York, NY: W.W. Norton & Company.

Nkrumah, Kwame. 1997. *Selected Speeches: Kwame Nkrumah*. Vols. 1–5. Compiled by Samuel Obeng. Accra, Ghana: Afram.

Nkrumah, Kwame. 1961. *I Speak of Freedom: A Statement of African Ideology*. Westport, CT: Paraeger Publishers, Greenwood Press.

Ntewesu, Samuel. 2005. *The Northern Factor in Accra: A Historical Study of Madina Zongo 1957-2000*. MA Thesis, University of Ghana.

Nyanzi, Stella, Ousman Bah, and Owen Rosenberg-Jallow. 2007. "Bumsters, Big Black Organs and Old White Gold: Embodied Racial Myths in Sexual Relationships of Gambian Beach Boys." *Culture, Health & Sexuality* 7 (6): 557–69.

O'Connell Davidson, Julia and Sanchez Taylor, J. 2005. "Travel and Taboo: Heterosexual Sex Tourism to the Caribbean." In Laurie Schaffner and Elizabeth Bernstein (eds.) *Regulating Sex: The Politics of Intimacy and Identity*. 83–100. London: Routledge.

O'Rourke, Harmony. 2012. "Native Foreigners and the Ambiguity of Order and Identity: The Case of African Diasporas and Islamic Law in British Cameroon." *History in Africa: A Journal of Method* 39 (2012): 97–122.

Odotei, Irene. 1972. *The Ga and Their Neighbors*. Ph. D. diss., University of Ghana, Legon.

Odotei, Irene. 1977. "The Ga-Damne." *Proceedings of the Seminar on Ghanaian Historiography and Historical Research/Ed. by J.o. Hunwick* P. 99–112.

Odotei, Irene. 1989. "What Is in a Name?: The Social and Historical Significance of Ga Names." *Research Review - Institute of African Studies* 5 (2): 34–51.

Odotei, Irene. 1991. "External Influences on Ga Society and Culture." *Research Review* 7 (1–2): 61–71.

Ofori-Ansa, Kwaku. 2003. "PANAFEST: A Potential Pot of Gold." *PANAFEST Foundation.* www.panafest.org

Ofori-Attah, Kwabena Dei. 2006. "The British and Curriculum Development in West Africa: A Historical Discourse." *International Review of Education* 52 (5): 409–23.

Olatunji, Babatunde, dir. 1998. *Love Drum Talk*. New York, NY: Chesky Records.

Olatunji, Babatunde, Robert Atkinson, and Akinsola A Akiwowo. 2005. *The Beat of My Drum: An Autobiography*. Philadelphia, PA: Temple University Press.

Olsen, Barbara. 1995. "Consuming Rastafari: Ethnographic Research in Context and Meaning." In Frank R. Kardes and Mita Sujan (eds.) *NA – Advances in Consumer Research*, Vol. 22, 481–85. Provo, UT: Association for Consumer Research.

Omolewa, Michael. 2006. "Educating The 'Native': A Study of the Education Adaptation Strategy in British Colonial Africa, 1910-1936." *Journal of African American History* 91 (3): 267–87.

Ong, Aihwa. 2006. *Neoliberalism as Exception: Mutations in Citizenship and Sovereignty*. Durham, NC: Duke University Press.

Opoku, A. A. 1970. *Festivals of Ghana*, Vol. 2. Accra: Ghana Publishing Corporation.

Oppong, Christine. 1974. "Attitudes to Family Size among Unmarried Junior Civil Servants in Acra." *Journal of Asian and African Studies* 9 (1–2): 76–82.

Osumare, Halifu. 2012. *The Hiplife in Ghana*. New York, NY: Palgrave Macmillan US.

Padilla, Mark, Jack Hirsch, Miguel Munoz-Laboy, Robert Sember, and Richard Parker, eds. 2007. *Love and Globalization: Transformations of Intimacy in the Contemporary World*. Nashville, TN: Vanderbilt University Press.

Parker, John. 2000. *Making the Town: Ga State and Society in Early Colonial Accra*. London: Pearson Education Ltd.

Parreñas, Rhacel Salazar. 2001. *Servants of Globalizatio : Women, Migration and Domestic Work*. Stanford, CA: Stanford University Press.

Pedersen, Morten Axel, and Morten Nielsen. 2013. "Trans-Temporal Hinges: Reflections on a Comparative Ethnographic Study of Chinese Infrastructural Projects in Mozambique and Mongolia." *Social Analysis: The International Journal of Social and Cultural Practice* 57 (1): 122–42.

Peil, Margaret. 1970. "The Apprenticeship System in Accra." *Africa* 40 (2): 137–50.

Pellow, Deborah. 1985. "Muslim Segmentation: Cohesion and Divisiveness in Accra." *The Journal of Modern African Studies* 23 (3): 419–44.

Pellow, Deborah. 2001. "Cultural Differences and Urban Spatial Forms: Elements of Boundedness in an Accra Community." *American Anthropologist* 103 (1): 59–75.

Pellow, Deborah. 2002. *Landlords and Lodgers: Socio-Spatial Organization in an Accra Community*. Westport, CT: Praeger.

Pellow, Deborah. 2003. "New Spaces in Accra: Transnational Houses." *City & Society* 15 (1): 59–86.

Perpener, John. 2001. *African-American Concert Dance: The Harlem Renaissance and Beyond*. Urbana, IL: University of Illinois Press.

Pessar, Patricia, and Sarah Mahler. 2001. "Gender and Transnational Migration." Transnational Communities Program Working Paper Series WPTC-01-20.

Phillips, Ruth B. 1994. "Fielding Culture: Dialogues between Art History and Anthropology." *Museum Anthropology* 18 (1): 39–46. https://doi.org/10.1525/mua.1994.18.1.39

Phillips, Ruth B. 2002. "Where Is 'Africa'? Re-Viewing Art and Artifact in the Age of Globalization." *American Anthropologist* 104 (3): 944–52. https://doi.org/10.1525/aa.2002.104.3.944

Phillips, Ruth B., and Christopher B. Steiner. 1999. *Unpacking Culture: Art and Commodity in Colonial and Postcolonial Worlds*. Berkeley, CA: University of California Press.

Pickard-Cambridge, Arthur Wallace. 1940. "The Place of Achimota in West African Education." *Journal of the Royal African Society* 39 (155): 143–53.

Pierre, Jemima. 2012. *The Predicament of Blackness: Postcolonial Ghana and the Politics of Race*. Chicago, IL: University of Chicago Press.

Pietz, William. 1985. "The Problem of the Fetish, I." *Res: Anthropology and Aesthetics* 9: 5–17.

Piot, Charles. 2010. *Nostalgia for the Future: West Africa after the Cold War*. Chicago, IL: University of Chicago Press.

Pogucki, R. J. H. 1954. *Report on Land Tenure in Customary Law of the Non-Akan Areas of the Gold Coast (Now Eastern Region of Ghana). Volume II*. Accra: Lands Department.

Polak, Rainer. 2000. "A Musical Instrument Travels Around the World: Jembe Playing in Bamako, West Africa, and Beyond." *The World of Music* 42 (3): 134–70.

Polak, Rainer. 2007. "Performing Audience: On the Social Constitution of Focused Interaction at Celebrations in Mali." *Anthropos* 102 (1): 3–18.

Povinelli, Elizabeth A. 2011. *Economies of Abandonment: Social Belonging and Endurance in Late Liberalism*. Durham, NC: Duke University Press.

Pratt, Mary Louise. 1992. *Imperial Eyes: Travel Writing and Transculturation*. London; New York, NY: Routledge.

Price, Sally. 1989. *Primitive Art in Civilized Places*. Chicago, IL: University of Chicago Press.

Pruitt, Deborah, and Suzanne LaFont. 1995. "For Love and Money: Romance Tourism in Jamaica." *Annals of Tourism Research* 22 (2): 422–40.

Quarcoopome, Samuel S. 1993. "A History of Urban Development of Accra: 1877–1957." *Research Review* 9 (1–2): 20–32.

Quarcoopome, Samuel S. 1998. "Social Impact of Urbanisation: The Case of Ga Mashie of Accra." *Transactions of the Historical Society of Ghana* 2 (2): 133–46.

Quaye, Irene. 1972. The Ga and their neighbours 1600-1742. Ph.D. History. University of Ghana.

Quayson, Ato. 2014. *Oxford Street, Accra: City Life and the Itineraries of Transnationalism*. Durham, NC: Duke University Press.

Ralph, Laurence. 2008. "Killing Time." *Social Text* 26 (4): 1–29.

Ray, Carina E. 2015. *Crossing the Color Line: Race, Sex, and the Contested Politics of Colonialism in Ghana*. New African Histories. Athens, OH: Ohio University Press.

Reed, Ann. 2014. *Pilgrimage Tourism of Diaspora Africans to Ghana*. New York, NY: Routledge.

Reed, Daniel B. 2016. *Abidjan USA: Music, Dance, and Mobility in the Lives of Four Ivorian Immigrants*. Reprint ed. Bloomington, IN; Indianapolis, IN: Indiana University Press.

Renne, Elisha P. 1995. *Cloth That Does Not Die: The Meaning of Cloth in Bùnú Social Life*. Seattle, WA: University of Washington Press.

Robertson, Claire C. 1984. *Sharing the Same Bowl? A Socioeconomic History of Women and Class in Accra, Ghana*. Bloomington, IN: Indiana University Press.

Roitman, Janet. 2013. *Anti-Crisis*. Durham, NC: Duke University Press.

Rosenbaum, Susanna. 2017. *Domestic Economies: Women, Work, and the American Dream in Los Angeles*. Durham, NC: Duke University Press.

Rosenbaum, Susanna and Ruti Talmor. 2022. "Self-Care." Keywords: A Feminist Vocabulary (Special Issue). *Feminist Anthropology* 3 (1): 362–372.

Ross, Mariama. 2000. Rasta Hair, US And Ghana: A Personal Note. In Sieber, Roy, Frank Herreman, Niangi Batulukisi, and N.Y. Museum for African Art, *Hair in African Art and Culture*. New York; Munich: Museum for African Art; Prestel.

Rouch, Jean. 1967. *Jaguar*. Paris: Les Films de la Pléiade.

Rovine, Victoria. 2015. *African Fashion, Global Style: Histories, Innovations, and Ideas You Can Wear*. Bloomington, IN: Indiana University Press.

Rovine, Victoria, and Susan Adams. 2002. "The Cultured Body." *African Arts* 35 (4): 1–91.

Rutherford, Danilyn. 2016. "Affect Theory and the Empirical." *Annual Review of Anthropology* 45: 285–300.

Ryan, Chris, and Colin Michael Hall. 2005. *Sex Tourism: Marginal People and Liminalities*. 1st ed. Hoboken, NJ: Taylor and Francis.

Sabatier, Peggy Roark. 1977. *Educating a Colonial Elite: The William Ponty School and Its Graduates*. Chicago, IL: The University of Chicago.

Sackeyfio, Naaborko. 2012. "The Politics of Land and Urban Space in Colonial Accra." *History in Africa* 39: 293–329.

Sadler, Michael E., ed. 1935. *Arts of West Africa (excluding music)*. London: International Institute of African Languages and Cultures, Oxford University Press, H. Milford.

Sahlins, Marshall D. 1963. "Poor Man, Rich Man, Big-Man, Chief: Political Types in Melanesia and Polynesia." *Comparative Studies in Society and History* 5 (3): 285–303.

Salazar, Noel B. 2010. *Envisioning Eden: Mobilizing Imaginaries in Tourism and Beyond*. New York, NY: Berghahn Books.

Salazar, Noel B., and Kiran Jayaram. 2016. *Keywords of Mobility: Critical Engagements*. New York, NY: Berghahn Books.

Salazar, Noel B, and Nelson H. H Graburn, eds. 2014. *Tourism Imaginaries: Anthropological Approaches*. New York, NY: Berghahn Books.

Sánchez-Taylor, Jacqueline. 2001. "Dollars Are a Girl's Best Friend? Female Tourists' Sexual Behaviour in the Caribbean." *Sociology* 35 (3): 749–64.

Sandelowsky, Beatrice H. 1976. "Functional and Tourist Art along the Okavango River." *Ethnic and Tourist Arts: Cultural Expressions from the Fourth World/Edited by Nelson H. Graburn*, 350–365.

Sapir, Edward. 1949. *Language: An Introduction to the Study of Speech. Harvest Books, Hb7*. New York, NY: Harcourt, Brace.

Sassen, Saskia. 1999. *Globalization and Its Discontents*. 1st ed. New York, NY: The New Press.

Saucier, P. Khalil, ed. 2011. *Native Tongues: An African Hip-Hop Reader*. Trenton, NJ: Africa World Press.

Savishinsky, Neil. 1994a. "Rastafari in the Promised Land: The Spread of a Jamaican Socioreligious Movement among the Youth of West Africa." *African Studies Review* 37 (3): 19–50.

Savishinsky, Neil. 1994b. "Transnational Popular Culture and the Global Spread of the Jamaican Rastafarian Movement." *Nwig: New West Indian Guide/Nieuwe West-Indische Gids* 68 (3/4): 259–81.

Sawyer, Lena. 2006. "Racialization, Gender and the Negotiation of Power in Stockholm's African Dance Courses." In Kamari Maxine Clarke and Deborah A. Thomas (eds.) *Globalization and Race: Transformations in the Cultural Production of Blackness*, 317–334. Durham, NC: Duke University Press.

Schauert, Paul. 2015. *Staging Ghana: Artistry and Nationalism in State Dance Ensembles*. Bloomington, IN; Indianapolis, IN: Indiana University Press.

Schildkrout, Enid. 1978. *People of the Zongo: the Transformation of Ethnic Identities in Ghana*. Cambridge; New York, NY: Cambridge University Press.

Schildkrout, Enid. 1998. "Africa: Permanent Installation." *African Arts* 30(1): 72–73.

Schildkrout, Enid, and Curtis A Keim, eds. 1998. *The Scramble for Art in Central Africa*. Cambridge: Cambridge University Press.

Schoss, Johanna. 1996. "Dressed To 'Shine': Work, Style and Leisure in Maundi, Kenya." In Hildi Hendrickson (ed.) *Clothing and Difference: Embodied Identities in Colonial and Post-Colonial Africa*, 156–88. Durham, NC: Duke University Press.

Seid'ou, Kari'kacha. 2014a. "Gold Coast Hand and Eye Work: A Genealogical History." *Global Advanced Research Journal of History, Political Science and International Relations* 3 (1): 8–16.

Seid'ou, Kari'kacha. 2014b. "Adaptive Art Education in Achimota College: G. A. Stevens, H. V. Meyerowitz and Colonial False Dichotomies." *CASS Journal of Art and Humanities* 3(1): 1–28.

Sennett, Richard. 1998. *The Corrosion of Character: The Personal Consequences of Work in the New Capitalism*. New York, NY: W.W. Norton.

Shankar, Shalini. 2008. *Desi Land: Teen Culture, Class, and Success in Silicon Valley*. Durham, NC: Duke University Press.

Sharpe, Jenny, and Suzanne Pinto. 2006. "The Sweetest Taboo: Studies of Caribbean Sexualities: A Review Essay." *Signs* 32 (1): 247–74.

Sheller, Mimi, and John Urry. 2004. *Tourism Mobilities: Places to Play, Places in Play*. London; New York, NY: Routledge.

Sherwood, Marika. 1996. *Kwame Nkrumah: the Years Abroad, 1935–1947*. Legon: Freedom Publications.

Shipley, Jesse Weaver. 2013. *Living the Hiplife: Celebrity and Entrepreneurship in Ghanaian Popular Music*. Durham, NC: Duke University Press.

Shipley, Jesse Weaver. 2015. *Trickster Theatre: The Poetics of Freedom in Urban Africa*. Bloomington, IN; Indianapolis, IN: Indiana University Press.

Simone, Abdoumaliq. 2009. *City Life from Jakarta to Dakar: Movements at the Crossroads*. New York, NY; London: Routledge.

Simpson, Kate. 2004. "'Doing development': The Gap Year, Volunteer-Tourists and a Popular Practice of Development." *Journal of International Development: The Journal of the Development Studies Association* 16 (5): 681–92.

Sin, Harng Luh. 2009. "Volunteer Tourism—"involve Me and I Will Learn"?" *Annals of Tourism Research* 36 (3): 480–501.

Singerman, Diane. 2007. "Middle East Youth Initiative, Wolfensohn Center for Development, and Dubai School of Government." *The Economic Imperatives of Marriage: Emerging Practices and Identities among Youth in the Middle East. Middle East Youth Initiative Working Paper* 6. Washington, DC: Wolfensohn Center for Development at the Brookings Institution.

Slater, Robert. 1930. "Changing Problems of the Gold Coast." *Journal of the Royal African Society* 29 (117): 461–66.

Slater, Don, and Janet Kwami. 2005. *Embeddedness and Escape: Internet and Mobile Use as Poverty Reduction Strategies in Ghana*. Working Paper (Information Society Research Group (ISRG)); No. 4. Accra: Department of International Development.

"Slave Voyages." n.d. Accessed June 21, 2023. https://www.slavevoyages.org/

Smith, Valene L., ed. 1989. *Hosts and Guests: The Anthropology of Tourism*. 2nd ed. Philadelphia, PA: University of Pennsylvania Press.

Sommers, Marc. 2012. *Stuck: Rwandan Youth and the Struggle for Adulthood*. Studies in Security and International Affairs. Atlanta, GA: University of Georgia.

Spronk, Rachel. 2009. "Sex, Sexuality and Negotiating Africanness in Nairobi." Africa: *The Journal of the International African Institute* 79 (4): 500–19.

Stasik, Michael, Valerie Hänsch, and Daniel Mains. 2020. "Temporalities of Waiting in Africa." *Critical African Studies* 12 (1): 1–9.

Steiner, Christopher B. 1994. *African Art in Transit*. Cambridge; New York, NY: Cambridge University Press.

Steiner-Khamsi, Gita, and Heidi Oxholm Quist. 2000. "The Politics of Educational Borrowing: Reopening the Case of Achimota in British Ghana." *Comparative Education Review* 44 (3): 272–99.

Stephens, Matthew. 1998. "Babylon's 'Natural Mystic': The North American Music Industry, the Legend of Bob Marley, and the Incorporation of Transnationalism." *Cultural Studies* 12 (2): 139–67.

Stephens, Michelle Ann. 2014. *Skin Acts: Race, Psychoanalysis, and the Black Male Performer*. Durham, NC: Duke University Press.

Stevens, George Alexander 1930. "The Future of African Art, with Special Reference to Problems Arising in Gold Coast Colony." *Oversea Education* 3 (2): 150–60.

Stevens, George A. 1935. "Educational Significance of Indigenous African Art." In Michael E. Sadler, ed. *Arts of West Africa* (excluding music). London [New York]: International Institute of African Languages and Cultures. 13–19. Oxford University Press, Humphrey Milford.

Stokes, Martin. 1999. "Music, Travel and Tourism: An Afterword." *The World of Music* 41 (3): 141–55.

Stoler, Ann Laura. 1995. *Race and the Education of Desire: Foucault's History of Sexuality and the Colonial Order of Things*. Durham, NC: Duke University Press.

Stoler, Ann Laura. 2002. *Carnal Knowledge and Imperial Power: Race and the Intimate in Colonial Rule*. Berkeley, CA: University of California Press.

Stoller, Paul. 2002. *Money Has No Smell: The Africanization of New York City*. Chicago, IL; London: University of Chicago Press.

Sunkett, Mark. 1993. *Mandiani Drum and Dance: Djimbe Performance and Black Aesthetics from Africa to the New World*. Performance in World Music Series. Tempe, AZ: White Cliffs Media.

Swan, Eileadh. 2012. "'I'm Not a Tourist. I'm a volunteer': Tourism, Development and International Volunteerism in Ghana." In Beek, Walter E. A. van, and Annette M. Schmidt (eds.) *African Hosts and Their Guests: Cultural Dynamics of Tourism*, 239–55. Oxford: James Currey.

Taylor, Jacqueline Sanchez. 2004. Tourism and 'embodied' commodities: sex tourism in the Caribbean. In Clift, Stephen, and Simon Carter, eds. Tourism and Sex: Culture, Commerce and Coercion. Tourism, Leisure, and Recreation Series. 41–53. London: Pinter.

Taylor, Steven J. L. 2019. *Exiles, Entrepreneurs, and Educators: African Americans in Ghana*. Suny Series in African American Studies. Albany, NY: State University of New York Press.

Taylor, Timothy Dean. 1997. *Global Pop: World Music, World Markets*. New York, NY: Routledge.

Tronto, Joan C. 1993. *Moral Boundaries: A Political Argument for an Ethic of Care*. New York, NY: Routledge.

Turino, Thomas. 2000. *Nationalists, Cosmopolitans, and Popular Music in Zimbabwe*. Chicago Studies in Ethnomusicology. Chicago, IL: University of Chicago Press.

Turner, Terence S. 2012. "The Social Skin." *HAU: Journal of Ethnographic Theory* 2 (2): 486–504.

Turner, Victor. 1969. "Liminality and Communitas." In *The Ritual Process: Structure and Anti Structure*, 94–113, 125–30. Chicago, IL: Aldine Publishing.

Turner, Victor. 1974. "Liminal to Liminoid, In Play, Flow, and Ritual: An Essay in Comparative Symbology." *The Rice University Studies* 60 (3): 53–92.

Twine, France Winddance. 2010. *A White Side of Black Britain: Interracial Intimacy and Racial Literacy*. Durham, NC: Duke University Press.

Urry, John. 1990. *The Tourist Gaze: Leisure and Travel in Contemporary Societies*. London: Sage Publications.

Utas, Mats, and Nordiska Afrikainstitutet. 2012. *African Conflicts and Informal Power: Big Men and Networks*. Africa Now. London: Zed Books.

Vigh, Henrik. 2009. "Wayward Migration: On Imagined Futures and Technological Voids." *Ethnos* 74 (1): 91–109.

wa Thiong'o, Ngũgĩ. 1986. *Decolonising the Mind: The Politics of Language in African Literature*. London; Portsmouth, NH: J. Currey; Heinemann.

Wacquant, Loïc. 2007. *Urban Outcasts: A Comparative Sociology of Advanced Marginality*. Hoboken, NJ: Wiley Press.

Wacquant, Loïc. 2008. "The Militarization of Urban Marginality: Lessons from the Brazilian Metropolis." *International Political Sociology* 2 (1): 56–74.

Wagner, Ulla, and Bawa Yamba. 1986. "Going North and Getting Attached: The Case of the Gambians." *Ethnos* 51 (3–4): 199–222. https://doi.org/10.1080/00141844.1986.9981323

Wallbank, Walter T. 1935. "Achimota College and Educational Objectives in Africa." *The Journal of Negro Education* 4 (2): 230–45.

Wearing, Stephen, and Jess Ponting. 2009. "Breaking down the system: How volunteer tourism contributes to new ways of viewing commodified tourism." In Jamal, Tazim, and Mike Robinson, eds. *The Sage Handbook of Tourism Studies*. 254–268. Los Angeles: SAGE.

Weiner, Annette. 1992. *Inalienable Possessions: The Paradox of Keeping-While-Giving*. Berkeley, CA: University of California Press.

Weisbord, Robert G. 1973. *Ebony Kinship: Africa, Africans, and the Afro-American*. Westport, CT: Greenwood.

Weiss, Brad. 2004. *Producing African Futures: Ritual and Reproduction in a Neoliberal Age*. Leiden; Boston, MA: Brill.

Weiss, Brad. 2009. *Street Dreams and Hip Hop Barbershops: Global Fantasy in Urban Tanzania*. Bloomington, IN; Indianapolis, IN: Indiana University Press.

Werbner, Pnina. 2008. *Anthropology and the New Cosmopolitanism: Rooted, Feminist and Vernacular Perspectives*. Oxford; New York, NY: Berg.

White, Cynthia M. 2007. "Living In Zion." *Journal of Black Studies* 37 (5): 677–709.

White, Cynthia M. 2010. "Rastafarian Repatriates and the Negotiation of Place in Ghana." *Ethnology* 49 (4): 303–20.

Whitehead, Clive. 2003. "Oversea Education and British Colonial Education 1929–63." *History of Education* 32 (5): 561–75.

Whitney, Michael L., and David Hussey. 1994. *Bob Marley: Reggae King of the World.* 2nd ed. San Francisco, CA: Pomegranate Artbooks.

Williams, Bianca C. 2018. *The Pursuit of Happiness: Black Women, Diasporic Dreams, and the Politics of Emotional Transnationalism.* Online Access with Subscription: Duke University Press. Durham, NC: Duke University Press.

Wilburn, Kenneth. 2012. "Africa to the World! Nkrumah-era Philatelic Images of Emerging Ghana and Pan-Africanism, 1957–1966." *African Studies Quarterly* 13 (2): 23–54.

Williams, Justin. 2015. "The 'Rawlings Revolution' and Rediscovery of the African Diaspora in Ghana (1983–2015)." *African Studies* 74 (3): 366–87.

Williams, Raymond. 1977. *Marxism and Literature.* Oxford: Oxford University Press.

Willis, Deborah, and Carla Williams. 2002. *The Black Female Body: A Photographic History.* Philadelphia, PA: Temple University Press.

Winegar, Jessica. 2006. *Creative Reckonings: The Politics of Art and Culture in Contemporary Egypt.* Stanford, CA: Stanford University Press.

Wolf, Diane L. 2002. "There's No Place Like "home": Emotional Transnationalism and the Struggles of Second-Generation Filipinos." In Waters, Mary C, and Peggy Levitt (eds.) *The Changing Face of Home: The Transnational Lives of the Second Generation,* 255–94. Book Collections on Project Muse. New York, NY: Russell Sage Foundation.

Wolff, Norma H. 2004. "African Artisans and the Global Market: The Case of Ghanaian 'Fertility Dolls.'" *African Economic History* 32 (32): 123–41.

Xiang, Biao, and Johan Lindquist. 2014. "Migration Infrastructure." *International Migration Review* 48: 122–48.

Yamaga, Chisono. 2006. *Japanese Girl Meets Nepali Boy: Mutual Fantasy and Desire in "Asian" Vacationscapes of Nepal.* MA Thesis, University of Manitoba.

Zelizer, Viviana A. 2005. *The Purchase of Intimacy.* Princeton, NJ: Princeton University Press.

Zimmerman, Mary K., Jennifer S. Litt, and Christine E. Bose. 2006. *Global Dimensions of Gender and Carework.* Stanford, CA: Stanford University Press.

Index

For Product Safety Concerns and Information please contact our EU
representative GPSR@taylorandfrancis.com
Taylor & Francis Verlag GmbH, Kaufingerstraße 24, 80331 München, Germany